Praise from the E

"Very hands-on and comprehensive! Both the novice and the advanced user will find valuable tips and techniques in this well-written text on SAS programming in the pharmaceutical industry. Important topics such as CDISC, MedDRA, WHODrug, and SOPs are clearly presented with the SAS user in mind. The text is a virtual treasure chest of many delightful examples, ranging from performing many-to-many comparisons/joins and creating box-and-whisker plots, to using the SQL Pass-Through Facility to read Microsoft Excel data."

Dr. Jimmy Thomas Efird
Director, Biostatistics and Data Management Facility
John A. Burns School of Medicine

"A down-to-earth book on the practical aspects of performing clinical trial analyses in the pharmaceutical industry. Jack Shostak wastes no time in getting to important issues such as industry regulations and standards, data preparation and transformation, and acquiring data. His use of examples helps reduce a lot of the development time one spends on getting the correct syntax for statements and procedures for reporting results, like creating tables and listings, preparing graphics, and performing commonly used statistical analyses. Programmers working with clinical trial data in areas outside the pharmaceutical industry might find this book useful in expanding their SAS programming skills as well."

Robert Francis, Ph.D.
NOVA Research Company

"Jack has done a great job of giving the most experienced clinical programmer or an entry-level clinical programmer what we love to read—source code and new approaches to industry issues. This book covers SAS and industry information from a clinical trial programmer's 'need to know' perspective. Consider this book the Ultimate Guided Expedition for Clinical Trial Programmers. This will keep new folks 'out of trouble' and give the seasoned professional something new to consider."

Tim Moore
President, Quality Research Partner, Inc.

SAS Press

SAS® Programming
in the Pharmaceutical Industry

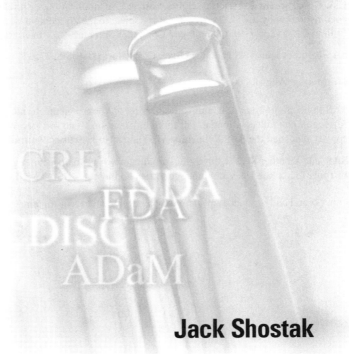

Jack Shostak

The Power to Know.

The correct bibliographic citation for this manual is as follows: Shostak, Jack. 2005. *SAS® Programming in the Pharmaceutical Industry*. Cary, NC: SAS Institute Inc.

SAS® Programming in the Pharmaceutical Industry

Contents

x

List of Programs

Programs in Chapter 5

Programs in Chapter 6

Programs in Chapter 8

Preface

This book was written for the entry- to intermediate-level SAS programmer who helps with the analysis and reporting of clinical trial data in the pharmaceutical industry. The industry may call this individual a "SAS programmer," "clinical SAS programmer," "statistical/stats programmer," "programmer/analyst," or any of a host of other names. In this book we call this individual a statistical programmer. Although this book may prove useful to clinical trial statisticians, it is aimed more at the statistical programmer who assists the lead trial statistician in producing the large amount of reporting required. This book assumes that the reader has no prior knowledge of clinical trials and some knowledge of SAS programming.

Most statistical programmers who end up working in clinical trial analysis happen upon the field by accident. This is primarily because historically there have been few if any formal training programs for clinical trial SAS programming. As a result, most statistical programmers have to learn about clinical trial analysis and reporting on the job. I thought it would be beneficial to have a resource for the junior statistical programmer that brings together in one volume much of the knowledge required to do clinical trial reporting.

This book is organized chronologically according to the statistical programmer's workflow. There is an introductory chapter that defines the working environment and sets the basic ground rules for the job. Then, there are chapters on importing data and massaging data into analysis data sets. Producing clinical trial report output is covered in the chapters on tables, listings, and graphs. Finally, there is a chapter on exporting data, followed by a discussion of the future for statistical programmers and a closing chapter on further resources.

The examples in this book focus primarily on the tools within SAS/GRAPH, SAS/STAT, and Base SAS, including the SAS macro language and PROC SQL. The examples were developed using SAS 9.1.3, but the vast majority of the examples will run with other versions of SAS as well. Please note that the data that drive the examples in this book are obtained through INPUT statements with DATALINES data. This is done only for illustrative purposes and does not mean that you should expect to obtain your clinical data in this fashion.

Acknowledgments

This book would not have been a success without the sage advice and hard work of a number of individuals. For steering this book in the right direction, I would like to thank these reviewers who truly shaped the content of this work: Tonya Balan, Nancy Cole, Chris Decker, Chris Holland, Michael Kilhullen, Angela Lightfoot, Gene Lightfoot, Pat Majcher, Carol Matthews, Hunter McGhee, Gina Marie Mondello, David Olaleye, Andy Ravenna, Paul Savarese, Brian Shilling, and Ron Vandenhouten.

Special thanks to the fine production team at SAS, who took my rough words and buffed and polished them into a book I am proud of. Thanks to Ed Huddleston, copyeditor, Candy Farrell, technical publishing specialist, and Patrice Cherry, designer, for making this a high-quality and attractive text. Thanks to Liz Villani and Shelly Goodin for helping me to get the word out about this book. Finally, I owe huge thanks to Judy Whatley, acquisitions editor, for sticking with me from day one on this effort. Thanks, Judy, for your perseverance, patience, and keen eye for detail, which helped me greatly.

Of course, thanks to all of you in the industry that I have worked with and learned from over the years—especially my friends and colleagues at the Duke Clinical Research Institute and my old friends at PRA and A.H. Robins.

Finally, thanks to my family, who were entirely supportive of me while I worked on this book. Linda, Chloe, and Alex—I owe you one. OK, I owe you a lot more than one.

Chapter 1

Environment and Guiding Principles

This chapter provides the context and universal guidelines for the material in this book. It is best to begin by describing the environment in which a statistical programmer works in the pharmaceutical industry. Then we explore the fundamental principles that should guide you in your day-to-day work. These principles permeate all of the tasks that you do on a daily basis and, if kept in mind, they will keep you from going astray in your statistical programming duties.

The Statistical Programmer's Working Environment

Pharmaceutical Industry Vocabulary

Like many industries, the pharmaceutical industry has a vocabulary and language all its own. Our industry is full of acronyms, medical terminology, and jargon that you must become familiar with to be an effective statistical programmer. To assist you in identifying some of these terms, this book *italicizes* the first occurrence of terminology specific to SAS programming in the pharmaceutical industry. At the end of the book is a glossary that you can refer to for definitions of these terms.

Statistical Programmer Work Description

The statistical programmer usually works in the statistics department of a pharmaceutical research and development group or *contract research organization* (*CRO*). The role of statistical programmers is to use their superior technical and programming skills in order to allow *clinical trial* statisticians to perform their statistical analysis duties more efficiently. This may involve importing and exporting data, working with other information technology professionals on site and at other companies, deriving variables and creating analysis data sets, and creating *clinical study report* (*CSR*) materials consisting of tables, figures, and listings (*TFL*). Here is a simplified illustration of the general work processes of the statistical programmer:

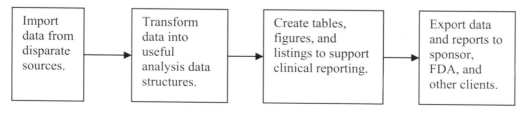

The Drug/Device Development Process

The clinical trial industry is primarily concerned with bringing new drugs, devices, or therapies to the general population. In the United States, most clinical trials are funded by pharmaceutical companies wanting to bring a new treatment to market or by the National Institutes of Health (NIH), which funds research to improve the health of all Americans. Because the majority of clinical trials are conducted with the idea to bring a new drug or device to market, we will briefly look at the U.S. *Food and Drug Administration* (*FDA*) approval process. Further details about the drug and device approval process can be found at www.fda.gov.

Drug Approval Process

The FDA is charged with making sure that all new drugs brought to market are both safe and effective. The FDA helps to do this with a drug approval process that can easily cost hundreds of millions of dollars and can take a decade or more to move a drug from discovery to a pharmacy near you. There are several progressive levels of studies that are conducted as part of the drug approval process.

1. **Pre-clinical studies** are the experiments that are conducted in the laboratory and with animals long before a new drug is ever introduced for use by humans. If these studies are promising, the drug maker usually pursues an *Investigational New Drug* (*IND*) application. The IND application allows the drug maker to conduct clinical trials of the new compound on human subjects.

2. **Phase 1 trials** are the "first in man" studies of a new drug in humans. These studies are usually carried out on small samples of subjects. The idea here is to determine the safety of the drug in a small and usually healthy volunteer study population.

3. **Phase 2 trials** go beyond phase 1 studies in that they begin to explore the efficacy of a drug. Phase 2 studies have larger (100–200 patients) study populations than phase 1 studies and are aimed at narrowing the dose range for the new medication. Safety is monitored at this stage as well, and phase 2 trials are generally conducted in the target study population.

4. **Phase 3 trials** are large-scale clinical trials on populations numbering in the hundreds to thousands of patients. These are the critical trials that the drug maker runs to show that its new drug is both safe and efficacious in the target study population. If the phase 3 trials are successful, they will form the keystone elements of a *New Drug Application* (*NDA*).

5. **Phase 4 trials**, or post-marketing trials, are usually conducted to monitor the long-term safety of a new drug after the drug is already available to consumers.

Device Approval Process

The FDA is also in charge of regulating new medical devices. The device approval process at the FDA varies based on the degree of risk inherent in the device. Class 1 devices carry little risk for the patient; they include devices such as elastic bandages and surgical instruments. Class 2 devices carry slightly higher risk for the patient; they include such devices as infusion pumps and motorized wheelchairs. Class 3 devices are high-risk devices and thus require the most regulatory scrutiny. Class 3 devices include replacement heart valves and implantable defibrillators. Obviously, the approval requirements for a class 3 device are much higher than for a class 1 device.

Clinical Trial Study Designs

There are many types of clinical trials, and there are some general trial design concepts that you need to understand. One key concept is the *randomization* of study therapy. When you randomly assign patients to study therapy, you reduce potential treatment *bias*. Another key concept is treatment *blinding*. Blinding a patient to treatment means the patient does not know what treatment is being administered. In a *single-blind trial*, only the patient does not know what treatment is being administered. In a *double-blind trial*, neither the patient nor the patient's doctor knows which treatment is being given. On occasion there may even be a *triple-blind trial*, where the patient, the patient's doctor, and the staff analyzing the trial data do not know what therapy is being given.

There are other trial design concepts for you to be aware of. A clinical trial can be carried out at a single *site* or it can be a *multi-center trial*. In a single-site trial all of the patients are seen at the same clinical site, and in a multi-center trial several clinical sites are used. Multi-center trials are needed sometimes to eliminate site-specific bias or because there are more patients required than a single site can *enroll*.

Trials may be designed to determine equivalence or superiority between therapies. An *equivalence trial* is designed to show that there is no clinically significant difference between therapies, and a *superiority trial* is intended to show that one therapy is significantly better than another.

Finally, trials can follow parallel or crossover study designs. In a *parallel trial*, patients are assigned to a therapy that they remain on, and they are compared with patients in alternate therapy groups. In a *crossover trial*, patients switch or change therapy assignments during the course of the trial.

Industry Regulations and Standards

Regulatory authorities govern and direct much of the work of the statistical programmer in the pharmaceutical industry. It is important for you to know about the following regulations, guidance, and standards organizations.

International Conference on Harmonization (ICH)

The *International Conference on Harmonization (ICH)* is a non-profit group that works with the pharmaceutical regulatory authorities in the United States, Europe, and Japan to develop common regulatory guidance for all three. The goal of the ICH is to define a common set of regulations so that a pharmaceutical regulatory application in one country can also be used in another. Over time the Food and Drug Administration (FDA) usually adopts the guidelines developed by the ICH, so you can watch the development of guidance at the ICH to see what FDA requirements may be forthcoming.

Clinical Data Interchange Standards Consortium (CDISC)

The *Clinical Data Interchange Standards Consortium (CDISC)* is a non-profit group that defines clinical data standards for the pharmaceutical industry. CDISC has developed numerous data models that you should familiarize yourself with. Four of these models are of particular importance to you:

- *Study Data Tabulation Model (SDTM)*. The SDTM defines the data tabulation data sets that are to be sent to the FDA as part of a regulatory submission. The FDA has endorsed the SDTM in its *Electronic Common Technical Document (eCTD)* guidance. The SDTM was originally designed to simplify the production of *case report tabulations (CRTs)*, and therefore the SDTM is listing friendly, but not necessarily friendly for creating statistical summaries and analysis.

- *Analysis Dataset Models (ADaM)*. The CDISC ADaM team defines data set definition guidance for the analysis data structures. These data sets are designed for creating statistical summaries and analysis.

- *Operational Data Model (ODM)*. The ODM is a powerful *XML*-based data model that allows for XML-based transmission of any data involved in the conduct of clinical trials. SAS has provided support for importing and exporting ODM files via the CDISC procedure and the XML LIBNAME engine.

- *Case Report Tabulation Data Definition Specification (Define.xml)*. Define.xml is the upcoming replacement for the data definition file (define.pdf) sent to the FDA with electronic submissions. Define.xml is based on the CDISC ODM model and is intended to provide a machine-readable version of define.pdf. Because define.xml is machine readable, the metadata about the submission data sets can be easily read by computer applications. This allows the FDA to work more easily with the data submitted to it.

You will be exporting, importing, and creating data for these models, so it is important that you learn about them. The FDA has begun to formally endorse the use of these data models in their guidance. Eventually the FDA will probably require data to be formatted to the CDISC model standards for regulatory submissions.

Food and Drug Administration (FDA) Regulation and Guidance

The FDA is the department within the United States Department of Health and Human Services that is charged with ensuring the safety and effectiveness of drugs, *biologics*, and devices marketed in the United States. Any work that you perform that contributes to a submission to the FDA is covered by these federal regulations. There are a number of specific regulations and guidance you must know.

"21 CFR – Part 11 Electronic Records; Electronic Signatures"

21 CFR – Part 11 is a federal law that regulates the submission of electronic records and electronic signatures to the FDA. Of particular interest to the statistical programmer are the following requirements of Part 11:

> "Validation of systems to ensure accuracy, reliability, consistent intended performance, and the ability to discern invalid or altered records."

> "Determination that persons who develop, maintain, or use electronic record/electronic signature systems have the education, training, and experience to perform their assigned tasks."

> "Adequate controls over the distribution of, access to, and use of documentation for systems operation and maintenance."

> "Revision and change control procedures to maintain an audit trail that documents time-sequenced development and modification of systems documentation."

21 CFR – Part 11 means that you must be qualified to do your work, your programming must be validated, you must have system security in place, and you must have change control procedures for your SAS programming. The current additional FDA guidance on 21 CFR – Part 11 is titled "Guidance for Industry Part 11, Electronic Records; Electronic Signatures—Scope and Application."

"E3 Structure and Content of Clinical Study Reports"

The "*E3*" describes in detail what reporting goes into a clinical study report for an FDA submission. This guidance is of major importance, as you are often required to generate tables, figures, case report tabulations, and perhaps *clinical narrative* support for the clinical study report.

"E9 Statistical Principles for Clinical Trials"

The "*E9*" discusses the statistical issues in the design and conduct of a clinical trial. It details trial design, trial conduct, and data analysis and reporting. Although most useful to the statistician, this guidance gives an excellent overview of how a clinical trial should be conducted.

"E6 Good Clinical Practice: Consolidated Guidance"

The "*E6*" (or *GCP*s) discusses the overall standards for implementing a clinical trial. Anyone who works on a clinical trial needs to understand this document. Of particular interest to the statistical programmer are the following parts of E6. The italics have been added for emphasis.

> "5.1.1 The sponsor is responsible for implementing and maintaining quality assurance and quality control systems with written SOPs [standard operating procedures] to ensure that trials are conducted and data are generated, documented (recorded), *and reported* in compliance with the protocol, GCP, and the applicable regulatory requirement(s)."

> "5.5.1 The sponsor should *utilize appropriately qualified individuals* to supervise the overall conduct of the trial, to handle the data, to verify the data, to *conduct the statistical analyses*, and to *prepare the trial reports.*"

> "5.5.4 *If data are transformed during processing, it should always be possible to compare the original data and observations with the processed data.*"

"Part 312.33 of Title 21 of the Code of Federal Regulations; Annual Reports"

21 CFR – Part 312.33 discusses what is required for an Investigational New Drug (IND) application. Part 312.33 discusses the requirements for the annual reporting for the IND. This reporting requires you to create adverse event, death, and subject dropout summaries annually for any drug under an IND application.

"Providing Regulatory Submissions in Electronic Format – General Considerations"

This guidance document governs how electronic files should be sent to the FDA. Currently, the FDA requests that electronic documents be submitted as Portable Document Format (PDF) files. The PDF page should be a standard 8.5" × 11" page with 1" margins and 12-point font. Data sets are currently to be sent to the FDA as SAS XPORT transport format files. In the future it is likely that data sets will be required to be sent as XML files, probably formatted in the CDISC ODM.

"Providing Regulatory Submissions in Electronic Format – NDAs"

This guidance document describes how a New Drug Application (NDA) may be sent electronically to the FDA. The guidance defines how the files in the electronic submission should be structured for FDA review.

"Electronic Common Technical Document Specification"

The Electronic Common Technical Document (eCTD) is the vision for future electronic submissions to the FDA. This specification was developed by the International Conference on Harmonization (ICH) as an open-standards solution for electronic submissions to worldwide regulatory authorities. The FDA has adopted the eCTD as the future replacement for its other e-submission guidance, although for now the older guidance is still in effect. Note that the eCTD still depends largely on submitting text documents as PDF files and submitting data sets as SAS XPORT transport format files.

Your Clinical Trial Colleagues

Within any pharmaceutical company or contract research organization, there are groups and individuals outside the statistics department that you work with. Let's take a look at the functional groups a statistical programmer interacts with most.

Site Management

The site management group is responsible for clinical site relations. They recruit doctors at clinics to participate in clinical trials, train their staff in trial conduct, and monitor the sites for protocol compliance while serving as an all-around advocate for the clinical site. Site management can be your ally in helping to get the data entered in a clean and readily usable form. Clean data at the start of the data collection process precludes the need for extra data queries from data management and helps prevent subsequent data analysis problems. With the arrival of *electronic data capture* (*EDC*) technology, the importance of site management has grown, because data entry has moved from the data management group to the clinical site itself.

Data Management

Next to the clinical trial statistician, the statistical programmer works most closely with the data management group. The data management group is usually responsible for *case report form* (*CRF*) design, database design and setup, data entry, data cleaning, data coding, data quality control, and providing the clinical trial data for analysis by the statistics group. Cleaning the data involves scouring the data for problems by using programmatic and manual checks of the data. Coding the data entails applying generic codes to freely entered text fields such as adverse events, medications, and medical histories. Quality control of the data involves auditing the data to make sure that it was entered properly. Finally, the data management group typically provides the data to the statistical programmer via some kind of *relational database management system* (*RDBMS*), which can then be imported into SAS. You save time when data management provides a well-cleaned and well-coded clinical database, because this means you do not have to program around dirty data.

Information Technology

The information technology (*IT*) group has varying responsibilities, depending on the size of your organization. IT is usually responsible for computer systems infrastructure, maintenance, and general computer help desk support. The IT group may also perform some level of software development. In small to midsize organizations IT may simply make *application program interfaces* (*APIs*) between off-the-shelf systems, while at large organizations IT may be responsible for full software applications architecture and development.

You need to work with the IT department within your organization as well as with external sponsors and vendors. Internally, you may work with IT for SAS configuration management and installation qualification, encryption technologies, and desktop publishing or report distribution concerns. The most common reason for you to work with external IT staff is usually in regard to information exchange technologies such as *FTP* and encryption tools.

Project Management

Most contract research organizations and pharmaceutical companies are organized in a matrix management structure. This structure is called a matrix because there are project teams that span various functional departments. It may help to visualize the relationship like this:

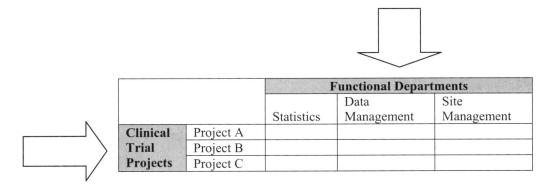

		Functional Departments		
		Statistics	Data Management	Site Management
Clinical Trial Projects	Project A			
	Project B			
	Project C			

The project manager provides operational oversight in a clinical trial. The project manager is responsible for meeting the trial needs by enlisting the support of the various functional departments. He or she also works with the primary investigator, as well as with external vendors such as laboratories, pharmaceutical companies, and contract research organizations. The project manager needs to work with the statistical programmer over the course of a clinical trial. As a statistical programmer, you may find that you answer to (at least) two managers during a trial. The statistical programming functional management serves as your skill-specific manager, while the project manager serves as your project-specific manager.

Quality Assurance

The quality assurance (QA) group is your internal regulatory reference, and they are there to help you. The primary goal of QA is to see that operations in your organization meet regulatory standards. They can assist you in interpreting the various regulations and help you to prepare for customer and regulatory audits. Quality assurance may also perform internal audits to make sure that your business processes meet regulatory standards. Finally, the QA group typically maintains all of your company's standard operating procedures (SOPs).

Medical Writing

The medical writing group may assist in creating various documents for your organization. Medical writers may help with the writing of clinical study reports for the FDA. Medical writers may also get involved in writing an NDA submission. Clinical narrative safety reporting is another task that medical writers help with. On occasion you will have to respond to requests for additional data from the medical writing group as they compile their reporting. Finally, a good medical writer can be a staunch ally in statistical reporting, as he or she may find any last-minute inconsistencies in your analysis before sending it along to the authorities.

Guiding Principles for the Statistical Programmer

The following are specific guiding principles for SAS programming in the pharmaceutical industry. These are high-level concepts that you should keep in mind while performing any of a broad range of tasks.

Understand the Clinical Study

A good statistical programmer takes time to understand the subject matter. If you were going to perform open-heart surgery and you were handy with a knife, you would not just roll up your sleeves and get to work. You would get formal training and obtain a medical degree first so that you understood what you were doing. The same can be said of SAS programming in the pharmaceutical industry. Just because you are a SAS expert does not mean you know all there is to know about a particular drug or device or the disease state it intends to cure.

There are several areas of study that will help you understand the research topic. As a first step you will want to read the clinical *protocol*. The protocol describes the device or medication to be used, the patient populations under study, the statistical plan of the clinical trial, and the details of the disease state. If you want to understand the disease state or indication further, you may want to seek out a clinical *investigator* of the clinical trial or do some further reading about the disease. Understanding the patient population is a good way to understand the data that you will see and whether there is reason for concern when viewing the data. For example, if you were studying a medication to reduce hypertension, you would not be as worried if patient blood pressure data were elevated at *baseline*. You would expect to see this because you understand that hypertensive patients have high blood pressure.

The next step in understanding the topic of study is to read the *statistical analysis plan* (*SAP*). The SAP is a very detailed document, separate from the protocol, describing how the clinical trial data will be analyzed. Although the protocol usually has only a few paragraphs on the statistical analysis, the SAP presents the entire statistical analysis in considerable detail. The SAP describes what inferential analyses will be done, defines the study population, presents data windowing or other special data handling rules, and often includes draft *output shells* that show precisely what tables, listings, and graphs will be provided in the reporting. The SAP is where the majority of your work is defined. Thus, you need to understand the SAP in exquisite detail, so it is beneficial to study it well in advance of programming.

There are additional documents describing the operations of the clinical trial that you may want to review. The site monitoring plan describes how the site management staff ensures that the clinical sites are conducting the protocol and completing the CRF properly. The clinical data management plan used by the data management staff may be useful to review. The clinical data management plan contains data entry instructions, data coding instructions, data review instructions, and a data quality control plan. Finally, there is the very important *annotated CRF,* which shows you where the variables in the clinical database come from on the CRF. The following is an example of an annotated medical history CRF page:

<data set name = MEDHX>

Protocol Name *<study>*   Patient Number: _ _ _ - _ _ _*<patid>*   Visit Identifier *<visit>*

Medical History:

	Historical Condition	**Diagnosis Date**
1	_____	_/_/___
2	_____	_/_/___
3	_____	_/_/___
4	_____	_/_/___

<seq>   *<condition>*   *<diagdt>*

Note that in this example the data set and variable names are in italics and are enclosed in angle brackets (< >). Also note that there may be external data (from the laboratory, ECG Holter monitor, etc.) loaded into your clinical data management system, and you will want the specifications for those data as well.

Program a Task Once and Reuse Your Code Everywhere

One of the main reasons that you use computers is to perform repetitive tasks for you. If you apply that line of reasoning to your statistical programming, it will serve you well by preventing you from "reinventing the wheel." Another way to describe this is that you should strive to make your statistical programming modular in nature. We can look at a demonstration of modular programming by examining the SAS libref. Many if not most SAS programs begin with SAS LIBNAME statements that look something like these.

Program 1.1 Librefs That Commonly Appear at the Start of a SAS Program

```
libname trialdata "c:\mytrial\sasdata";
libname library "c:\mytrial\mysasformats";
libname otherdata "c:\someotherdata";
```

For a clinical trial you may have anywhere from a few dozen to a couple hundred different SAS programs, depending on the nature of the project. If you were to copy the LIBNAME statements above into 200 different programs, you might realize two things:

1. You had to copy three common lines of SAS code into 200 different places.
2. You now have a code maintenance problem.

The code maintenance problem surfaces when you realize that you need to change one of those SAS librefs. Then you have to edit many SAS programs to make this simple change. An alternative to having those three SAS librefs everywhere is to have them in a single location. The SAS macro facility provides two simple ways to do this. You could place those three LIBNAME statements in a single SAS program and use a %INCLUDE macro statement like this:

```
%include "c:\mylibrefs.sas";
```

Another approach would be to wrap a SAS macro around the three LIBNAME statements and call it with a simple SAS macro call. First, set up a SAS macro.

Program 1.2 Using a SAS Macro to Define Common Librefs

```
%macro mylibs;
    libname trialdata "c:\mytrial\sasdata";
    libname library "c:\mytrial\mysasformats";
    libname otherdata "c:\someotherdata";
%mend mylibs;
```

Then, call the SAS macro in another SAS program like this:

```
%mylibs
```

With either approach, what you have done is take a piece of SAS code common to many programs and put it in one place. If you ever have to make a change to one of those SAS librefs across all programs, you can easily change it in a single place. This practice is fundamental to good programming, and although it is possible to be overly modular, it is better to err on the side of making your SAS code more modular than to create SAS code maintenance problems over the long term.

Clinical Trial Data Are Dirty

People and their behavior are unique, and that is a wonderful thing. Unfortunately, the data that describe a patient's activities during a clinical trial tend to be unique as well. The clinical trial protocol and clinical trial staff make the best effort to guide the patient through a common treatment protocol, but this is often not enough to control the data coming from the patient. It is also often the case that the case report form used to collect the data turns out to be a less than perfect instrument for collecting what is needed for analyses. Finally, despite the best efforts of data management to provide a clean database, not all data fields are scoured. Therefore, you may be faced with a sometimes deviant and heterogeneous clinical trial database, so you need to be on guard for dirty or discrepant data.

The best way to protect yourself against dirty clinical trial data is to use good defensive programming techniques. In other words, you should write SAS code that accounts for all possible data permutations. Imagine you have a SAS data set that contains adverse event data for patients in a trial. Assume that the data set has only three fields: the subject ID (subjectid), a "yes or no" field describing whether the subject had an adverse event or not (aeyn), and the text description of the adverse event (aetext). To extract data for the patients who had an adverse event, you might set up a SAS data set as in the following program.

Program 1.3 Subsetting a Data Set for Patients with an Adverse Event

```
data aes;
   set aes;
      by subjectid;
         where aeyn = "YES";
run;
```

Now, consider what would happen if the SAS data set "aes" looks like this:

subjectid	aeyn	aetext
101	YES	Rash
102		Hives
103	NO	
104	YES	Headache

The SAS code you wrote would eliminate the observation for subjectid=102. This is because the "aeyn" field is not populated for that row and is therefore eliminated by the WHERE clause in SAS. This is a classic "*parent-child*" data problem in clinical trial data, where the "parent" question is left unanswered but the "child" response is given. A way to handle this problem would be either to include the "aetext" field in the WHERE clause or to add a warning to the SAS log. The code in Program 1.4 does both.

Program 1.4 Subsetting a Data Set for Patients with an Adverse Event Using Defensive Programming Techniques

```
data aes;
   set aes;
      by subjectid;

      **** PARENT-CHILD WARNING;
      if (aeyn ne "YES" and aetext ne "") or
         (aeyn = "YES" and aetext = "") then
            put "WARN" "ING: ae parent-child bug " aeyn= aetext=;
```

```
        **** GET AES;
        if aeyn = "YES" or aetext ne "";
run;
```

This SAS program first warns you when the "parent-child" data fields are out of synchrony and subsequently keeps all observations that could possibly indicate an adverse event. (Note that when the PUT statement is triggered, the "WARN" and "ING" are concatenated in the log file and signal a warning condition to SAS. The same trick can be used with "ERR" "OR" conditions as well. The benefit of breaking the "WARNING" and "ERROR" text in half in the SAS program is that it will be missed during text searching of SAS log files for warning and error conditions if none exist.)

Anywhere you have conditional logic is another place for defensive programming techniques. When there is conditional logic, there should be a catch-all follow-up statement. Assume you have SAS code such as the following.

Program 1.5 Example of Simple Conditional Logic IF-THEN/ELSE

```
if a > b then
    a + b;
else if a < b then
    a - b;
```

There should always be a follow-up ELSE statement to trap any potentially unforeseen conditions like the following.

Program 1.6 Example of Simple Conditional Logic IF-THEN/ELSE Using Defensive Programming Techniques

```
if a > b then
    a = a + b;
else if a < b then
    a = a - b;
else
    put "How does a relate to b? " a= b=;
```

The SAS SELECT statement is great for conditional processing because it has a mandatory OTHERWISE clause built into it to help catch unforeseen conditions. In a SAS SELECT statement, the code above would look like the code in the following program.

Program 1.7 Example of Simple Conditional Logic SELECT

```
select;
   when(a > b) a = a + b;
   when(a < b) a = a - b;
   otherwise put "What am I missing? " a= b=;
end;
```

In an optimal world, the CRF is perfectly designed to answer the questions of the study and the clinical data management group will have cleaned the data to perfection. However, to be a good statistical programmer in the clinical trial arena, you must always keep a lookout for errant data and program defensively. Defensive programming lets you account for all possible clinical data permutations.

Use SAS Macros Judiciously

The SAS macro language is a very powerful tool. With SAS macros you can write dynamic SAS applications that are in essence SAS programs that write other SAS programs. Unfortunately, with such great power comes the potential for great abuse. The SAS macro language can be abused when it is used to such an extent that a SAS program becomes unreadable. A SAS macro can become unreadable when it is too dense with macro invocations, is poorly documented, or involves too many nested macro calls. For instance, examine the following SAS code.

Program 1.8 Example of SAS Macro Code That You Should Not Write

```
data &&some&i;
   &getfile &subopt;
      &subset;
      %makecod
%run
```

Perhaps this is valid SAS code, but there is no code documentation to tell the user what any of those macro variables or macro calls actually do. Upon further investigation, we may find that the %MAKECOD macro calls six other SAS macros in a nested fashion. Also, is a %RUN really necessary, or has the programmer developed one too many macros?

The SAS macro language can also be abused when it is used in place of a built-in facility of SAS designed to solve the given task. A classic example is using the SAS macro language when simple SAS BY statement processing would work in its place. Examine the following SAS code, which prints out demographic data patient by patient.

Program 1.9 Reinventing SAS BY Processing with a SAS Macro

```
proc sort
   data = demog;
      by subjectid;
run;

**** SAS MACRO TO PRINT MY DEMOGRAPHIC DATA BY PATIENT;
%macro printpt(subjectid);
   proc print
      data = demog;
      where subjectid = "&subjectid";
      var subjectid age sex race;
   run;
%mend printpt;

%printpt(101-001)
%printpt(101-002)
%printpt(101-003)
```

Although the SAS code is valid and gets the job done, the following SAS code is better because it can handle unlimited "subjectid" while at the same time being less cumbersome to read.

Program 1.10 Using SAS BY Processing Instead of a SAS Macro

```
proc sort
   data = demog;
      by subjectid;
run;

**** PRINT MY DEMOGRAPHIC DATA BY PATIENT;
proc print
   data = demog;
      by subjectid;
         var subjectid age sex race;
run;
```

When the SAS macro language is used judiciously, it can be a powerful ally. SAS macros should be written so that they are not overly complicated, and they should always be the best-documented SAS programming code in any application. There is no worse fate than to be handed a complex SAS macro program with insufficient documentation. For more information about sound SAS macro programming practices, you can refer to the SAS Press books *Carpenter's Complete Guide to the SAS Macro Language*, by Art Carpenter, and *SAS Macro Programming Made Easy*, by Michele Burlew.

A Good Programmer Is a Good Student

Anyone who has programmed in any language has discovered that one of the best ways to learn to program is to learn from the examples of others. We all innovate when we author programs, but when faced with a new problem it is natural to look to others to see how they may have approached it. When you look to others for SAS programming help you will pick up new SAS tricks and insights. Each insight you gain from studying the efforts of others can be extrapolated into creating your own programs. This is essential to becoming a good programmer. So, where can you find a teacher or mentor?

Your search for the knowledge of SAS programming should begin locally. Look around in your department or organization for other statistical programmers who might serve as mentors and resources. These sage individuals have much to offer and can provide hands-on help to you in enhancing your programming skill. There is a wealth of knowledge outside your organization as well. SAS and SAS Publishing provide many papers and books that show how to maximize your use of SAS software. Local and national SAS users groups are also an excellent source of published papers on how to innovate in SAS. Finally, the Internet contains a treasure trove of SAS tricks and tips. There are many individual Web sites devoted to SAS, as well as the SAS-L listserv, which is a major source of SAS help. If you have a question related to SAS, it has probably been asked on SAS-L at some point. If it has not already been asked, then someone reading SAS-L will likely have an answer for you. Chapter 10, "Further Resources," contains a section that explains in detail how to use these resources.

Strive to Make Your Programming Readable

It is important that you make your programs readable. It is guaranteed that someone other than you will have to read your program at some point, and you want to make that task easy. A well-written SAS program has an ample supply of comments, employs white space to make the code pleasing to the eye, and uses a consistent case and indention style. For more information about basic SAS programming style, you may want to refer to *The Little SAS Book: A Primer*, by Lora Delwiche and Susan Slaughter.

Chapter 2

Preparing and Classifying Clinical Trial Data

This chapter describes the key clinical data preparation issues and the different classes of clinical data found in clinical trials. Each class of data brings with it a different set of challenges and special handling issues. Sample case report form (CRF) pages are provided with each type of data to aid you in visualizing what the data look like. The key data preparation issues presented are concepts that apply universally across the various classes of clinical trial data.

Preparing Clinical Trial Data

Clinical trial data come to the statistical programmer in two basic forms: numeric variables and character string (text) variables. With this in mind, there are two considerations for all numeric and text variables. All data should be cleaned if they are needed for analyses, and any data entered as *free-text variables* should be coded or categorized if they are needed for analyses.

"Clean" the Data If They Are Needed for Analysis

If data will be summarized or analyzed as part of the protocol-defined statistical analysis, they should be cleaned first. "Cleaned" in this context means that erroneous data entered into a variable are repaired before data analysis. Under the direction of the statistics group, the data management group is responsible for cleaning the clinical data.

Before the statistical programmer receives data that are ready for analysis, the clinical data management group cleans the data. This is done through a query process, which is built into the clinical data management system. The clinical data management query process usually looks like this:

1. A programmatic or manual investigation finds an errant data point.
2. A "query" or data clarification form (DCF) for that data point is sent to the clinical site.
3. The clinical site responds to the query.
4. The clinical data management group updates the database or CRF based on the response from the clinical site.

Depending on the size and complexity of the clinical trial, queries sent to sites can easily number into the thousands. Because the cost of reconciling these queries quickly rises, it is important to be judicious in creating them. It is worth noting that electronic data capture (EDC) systems reduce the number of queries needed, because the entry screens are often programmed so that errant data cannot be entered.

In order to reduce unnecessary data queries, the statistics group should be consulted early in the clinical database development process to identify variables critical for data analysis. Optimally, the statistical analysis plan would already be written by the time of database development so that the queries could be designed based on the critical variables indicated in the analysis plan. However, at the database development stage, usually only the clinical protocol exists to guide the statistics and clinical data management departments in developing the query or data management plan.

How clean the data must be depends on the importance of the data. Critical analysis variables must be clean, so this is where the site and data management groups should focus their resources. If the data are "dirty" at the time of statistical analysis, many inefficient and costly workarounds may need to be applied in the statistical programming, and the quality of the data analysis could suffer. However, if a variable is not important to the statistical analysis, then it is better to save the expense of cleaning that variable.

Categorize Data If Necessary

Clinical trial data come in two basic forms: numeric variables and text variables. Numeric variables are easy for the statistical programmer to handle. Numbers can be analyzed with SAS in a continuous or categorical fashion without much effort. If a numeric variable needs categorization, it is easy enough to categorize the data within SAS. For example, if you had to classify patient age, a simple DATA step such as the following might serve well.

Program 2.1 Categorizing Numeric Data

```
data demog;
   set demog;
      by subject;

      if .z < age <= 18 then
         age_cat = 1;
      else if 18 < age <= 60 then
         age_cat = 2;
      else if 60 < age then
         age_cat = 3;
   run;
```

The problem for the statistical programmer in categorizing data comes from text variables, or more specifically, free-text variables. A "free-text" variable is one that may contain any characters and is typically limited only in length. As an example, let's say you need to summarize the adverse events for a set of patients in a trial. The following SAS code shows the data and a quick summarization of the adverse events.

Program 2.2 Summarizing Free-Text Adverse Event Data

```
data adverse;
   input subjectid $ 1-7 ae $ 9-41;
   datalines;
   100-101 HEDACHE
   100-105 HEADACHE
   100-110 MYOCARDIAL INFARCTION
   200-004 MI
   300-023 BROKEN LEG
   400-010 HIVES
   500-001 LIGHTHEADEDNESS/FACIAL LACERATION
   ;
run;

proc freq
   data = adverse;
   tables ae;
run;
```

Program 2.2 yields the following output.

```
The FREQ Procedure


                                              Cumulative   Cumulative
ae                         Frequency  Percent  Frequency     Percent
_____

BROKEN LEG                         1    14.29          1       14.29
HEADACHE                           1    14.29          2       28.57
HEDACHE                            1    14.29          3       42.86
HIVES                              1    14.29          4       57.14
LIGHTHEADEDNESS/FACIAL LACERATION  1    14.29          5       71.43
MI                                 1    14.29          6       85.71
MYOCARDIAL INFARCTION              1    14.29          7      100.00
```

There are three problems with this adverse events summary. First, "HEADACHE" and "HEDACHE" are counted as separate events even though it is clear that the latter is simply a misspelling of the former. Second, "MI" and "MYOCARDIAL INFARCTION" are considered as separate events even though the former is simply an abbreviation of the latter. Finally, "LIGHTHEADEDNESS/FACIAL LACERATION" refers to perhaps related but different adverse events that need to be counted separately. All three of these problems exist because the data were entered in free-text fashion.

There is only one good solution to handling free-text variables that are needed for statistical analysis. The free-text variables need to be coded by clinical data management in the clinical database. If the adverse events were coded with a dictionary such as *MedDRA*, the previous example might look like Program 2.3.

Program 2.3 Summarizing Coded Adverse Event Data

```
data adverse;

    label aecode      = "Adverse Event Code"
          ae_verbatim = "AE Verbatim CRF text"
          ae_pt       = "AE preferred term";

    input subjectid $ 1-7 aecode $ 9-16
          ae_verbatim $ 18-39 ae_pt $ 40-60;

    datalines;
    100-101 10019211 HEDACHE                HEADACHE
    100-105 10019211 HEADACHE               HEADACHE
    100-110 10028596 MYOCARDIAL INFARCTION MYOCARDIAL INFARCTION
    200-004 10028596 MI                     MYOCARDIAL INFARCTION
    300-023 10061599 BROKEN LEG             LOWER LIMB FRACTURE
    400-010 10046735 HIVES                  URTICARIA
    500-001 10013573 LIGHTHEADEDNESS        DIZZINESS
    500-001 10058818 FACIAL LACERATION      SKIN LACERATION
    ;
run;

proc freq
   data = adverse;
   tables ae_pt;
run;
```

Program 2.3 yields the following output.

```
The FREQ Procedure


                            AE preferred term


                                            Cumulative     Cumulative
ae_pt                  Frequency    Percent   Frequency        Percent


DIZZINESS                      1      12.50           1          12.50
HEADACHE                       2      25.00           3          37.50
LOWER LIMB FRACTURE            1      12.50           4          50.00
MYOCARDIAL INFARCTION          2      25.00           6          75.00
SKIN LACERATION                1      12.50           7          87.50
URTICARIA                      1      12.50           8         100.00
```

You can see the benefit of coding the adverse events in the resulting summary. The headaches and myocardial infarctions are grouped appropriately, and splitting lightheadedness and facial laceration into separate events leads to those data being summarized separately as well.

However, there are some alternative, albeit poor, solutions to the free-text variable problem. One option is to *hardcode* events so that they are categorized properly. We will discuss hardcoding further in the next section, but it is generally a practice to be avoided as much as possible. Another option is to use a SAS DATA step string function such as SOUNDEX, INDEX, INDEXW, or SUBSTR to try to categorize data in like groups. This approach is very risky, as you cannot be guaranteed to capture all free-text data and categorize them the same way with these text-scanning tools. If the free-text data are unimportant, then such tools can be used. However, if the data are unimportant, then they probably should not be analyzed anyway and at best should be presented in some type of data listing.

Avoid Hardcoding Data

Sometimes even after a good attempt by clinical data management at cleaning and coding the data you may find that the data still contain some undesired or discrepant values. Perhaps a variable was left uncoded, or perhaps there is a serious adverse event known to have occurred that has not yet been entered in the clinical database. When this happens, the statistical programmer may use hardcoding. Hardcoding is explicitly stating the value of a symbolic object or variable in a program. An example of hardcoding follows.

Program 2.4 A Hardcoding Example

```
data endstudy;
   set endstudy;

   if subjid = "101-1002" then
      discterm = "Death";
run;
```

In this example, it is known from non-database sources that at study termination, subject 101-1002 died. That information is hardcoded into the program and overrides the information coming from the clinical data management system. Here are two reasons why hardcoding is a bad practice:

- Hardcoding overrides the database controls in a clinical data management system. With hardcoding there is no clear audit trail of data change and CFR 21 – Part 11 controls might be considered compromised.

- Data often change in a trial over time, and the hardcode that is written today may not be valid in the future. Unfortunately, a hardcode may be forgotten and left in the SAS program, and that can lead to an incorrect database change.

Many organizations expressly forbid hardcoding in their SAS programming standard operating procedures, while others allow the practice. Occasionally, there may be a justifiable reason for hardcoding. For instance, there may be an upcoming *data safety and monitoring board* (*DSMB*) or *independent data monitoring committee* (*IDMC*) meeting where the clinical trial must be monitored for safety information using the best available data. If there is a critical adverse event that the statistical staff is aware of but it cannot be entered in the clinical data management system in time, then perhaps that would justify hardcoding. However, it is better to avoid hardcoding at all costs and instead correct data in the clinical data management system. If hardcoding must be done, then an approach like the following might be used.

Program 2.5 An Improved Hardcoding Example

```
data endstudy;
   set endstudy;

   **** HARDCODE APPROVED BY DR. NAME AT SPONSOR ON 02/02/2005;
   if subjid = "101-1002" and "&sysdate" <= "01MAY2005"d then
      do;
         discterm = "Death";
         put "Subject " subjid "hardcoded to termination reason"
             discterm;
run;
```

Note that this program uses SAS code comment text to indicate that hardcoding is being used and with whose approval. Requiring a keyword such as "HARDCODE" in the comment facilitates searches for hardcodes later. Also, note that a PUT statement is provided to the SAS log, verifying during program execution that hardcoding has been used. Finally, the hardcode in Program 2.5 "expires" after May 1, 2005. If you know that an IDMC meeting will be held in April 2005 and you do not want to worry about old hardcodes, you could program them so that they expire in this way.

In summary, for data to be useful in clinical trial analyses they need to be quantifiable. The data must be either a continuous measure or a categorical value. Free text poses a problem for analysis, and if it is a valuable variable for the statistical analyses it really must be coded. Finally, hardcoding should be used only when absolutely necessary, because it is inherently problematic. Organizations that do allow hardcoding should document in their standard operating procedures (SOPs) that it is an approved business practice and how it is to be used.

Classifying Clinical Trial Data

There are different ways to classify clinical trial data. As mentioned earlier, data can be classified by their physical nature into discrete chunks or as a more continuous measurable quantity. In clinical trials there are other important contextual ways of grouping data as well. For instance, clinical trials are primarily focused on determining two things about a drug, biologic, or device: Is it efficacious, and is it safe? The data that help to answer these questions are broadly classified as *efficacy data* and *safety data,* respectively.

The Clinical Data Interchange Standards Consortium (CDISC) and its Submission Data Standards group have provided another way to broadly categorize clinical trial data. They have categorized data into *interventions class, events class, findings class,* and other special-purpose "*domains*" such as demographics. Interventions are the drug administration and surgical procedures that the patient receives during the course of the trial. Events are the unplanned clinical occurrences that the patient experiences over the course of the trial. Findings capture the planned examinations of the patient over the course of the trial. The demographics of a patient are that person's essential baseline characteristics.

Demographics and Trial-Specific Baseline Data

Here is a typical demographics CRF:

Protocol Name	Patient Number: _ _ _ - _ _ _	Patient Initials: _ _ _

DEMOGRAPHICS:

Birth Date: _ _ / _ _ _ / _ _ _ _ (Day/Month/Year)
Gender: ☐ Male ☐ Female
Race: ☐ Caucasian ☐ Black ☐ Asian ☐ Other
Ethnicity: ☐ Hispanic ☐ Non-Hispanic

Trial-specific patient characteristics may be included with the demographics data as well. Height, weight, smoking status, and sometimes vital signs are common additions. These characteristics are collected because they may be relevant to the study topic and could be used to stratify the statistical analysis. Baseline demographic characteristics are used to define patient groupings, or *strata*, for subpopulation analyses, or they may be used as *covariates* during *inferential analyses*. Demographic and baseline characteristic data are also commonly used to show that the therapeutic treatments under study have comparable populations at baseline. Demographics data fall into the demographics CDISC domain and play a part in efficacy and safety analyses, as either may be stratified by demographic characteristics.

Concomitant or Prior Medication Data

Concomitant medications and *prior medications* are collected in one of two forms: a list-type free-text format where the medications get coded later by data management, or a pre-categorized data format. Here is the free-text CRF format:

Protocol Name	Patient Number: _ _ _ - _ _ _	Visit Identifier

Concomitant Medications:

	Medication	Daily Dose	Start Date	Stop Date	Reason
1	_____	____	_/_/___	_/_/___	_____
2	_____	____	_/_/___	_/_/___	_____
3	_____	____	_/_/___	_/_/___	_____
4	_____	____	_/_/___	_/_/___	_____
etc.					

Here is the pre-categorized CRF format:

| Protocol Name | Patient Number: _ _ _ - _ _ _ | Visit Identifier |

Concomitant Medications:

Medication	Yes	No	Start Date	Stop Date	Reason
ACE Inhibitor	—	—	_/_/___	_/_/___	_____
Anticonvulsant	—	—	_/_/___	_/_/___	_____
Beta Blocker	—	—	_/_/___	_/_/___	_____
Psychoactive Medication	—	—	_/_/___	_/_/___	_____
etc.					

The free-text CRF format is useful in that it allows for an explicit description of the medication taken, whereas the pre-categorized format omits that detail. However, the free-text list format necessitates additional coding with a coding dictionary such as *WHOdrug* in order to be useful for analyses. The pre-categorized format has the benefit of capturing only the medications of concern for the given therapy and eliminates the cost of additional coding.

An essential detail for the statistical programmer to watch for in prior or concomitant medications data is whether or not the start and stop dates are important for analyses. Unfortunately, it is often the case that the importance of the timing of prior or concomitant medications is not determined until after much of the data have been entered or even after the database is closed to entry. For instance, it may be decided later that a specific concomitant medication has to be watched carefully for interaction with a medication used in the study. If insufficient attention was placed on the quality of the medication start and stop dates, then determining whether there is overlap with study medication is difficult if not impossible.

Concomitant or prior medications may be used in either safety or efficacy analyses. The presence of specific medications may be used as covariates for inferential analyses. Also, medications are often summarized to show that the therapies under study come from medically comparable populations. Medications may be used to determine protocol compliance and to help define a protocol-compliant study population. Concomitant medications may be examined to determine whether they interact with study therapy or whether they can explain the presence of certain adverse events. From a CDISC perspective, prior medications would be considered a finding while concomitant medications would be considered an intervention.

Medical History Data

Like concomitant medication data, patient *medical history* data are collected in one of two forms: a list-type free-text format where the histories get coded, or a pre-categorized data format. Here is the free-text CRF format:

Protocol Name _____   Patient Number: _ _ _ - _ _ _   Visit Identifier _____

Medical History:

	Historical Condition	Diagnosis Date
1	_____	__/__/____
2	_____	__/__/____
3	_____	__/__/____
4	_____	__/__/____
	etc.	

Here is the pre-categorized CRF format:

Protocol Name _____   Patient Number: _ _ _ - _ _ _   Visit Identifier _____

Medical History:

Historical Condition	Yes	No	Diagnosis Date
Diabetes	__	__	__/__/____
Stroke	__	__	__/__/____
Hypertension	__	__	__/__/____
Neurological Disorders	__	__	__/__/____
etc.			

Again, the free-text CRF format is useful in that it allows for explicit description of the historical condition, whereas the pre-categorized CRF format omits that detail. However, the free-text list format necessitates coding with a coding dictionary such as MedDRA in order to be useful for analyses. The pre-categorized format is useful here, as only medical history relevant to the investigational therapy can be captured and the cost of additional coding of the history data is eliminated entirely.

Medical history data may be used in either safety or efficacy analyses. The presence of historical medical conditions may be used as covariates for inferential analyses. Also, medical histories are typically summarized to show that the therapies under study come from study populations with comparable disease histories. Medical histories may be used to determine protocol compliance and to help define a protocol-compliant study population. Medical history is considered a finding from a CDISC perspective.

Investigational Therapy Drug Log

Drug logs capture the investigational drug dosing times. Here is a sample drug log CRF form:

Protocol Name	Patient Number: _ _ _ - _ _ _
Study Drug Dosing:	

Dose #	Date of Dosing	Time of Dosing (24-hour clock)	Total Dose (mg)
1	_/_/_	_ : _	_____
2	_/_/_	_ : _	_____
3	_/_/_	_ : _	_____
4	_/_/_	_ : _	_____
etc.			

The investigational therapy drug log can be a source of problems for the statistical programmer. Here again, dates and times of dosing may be critical to effective use of this data. Missing dosing records, start times, or stop times can seriously hinder the quality of the reporting of dosing data. It is important to look at the analysis plan and determine if the dosing data are important to analysis. If they are important, then data management should clean the data to ensure the quality of the medication start and stop times.

Drug log or dosing data are used in many ways for both efficacy and safety analyses. As a safety issue, the drug record is often used in conjunction with adverse events to determine whether adverse events were treatment emergent. In other words, did the patient have an adverse event that might have been caused by the investigational therapy? Also, drug log data may be used for safety analysis purposes to watch for abnormal laboratory values or other clinical events after dosing. Finally, drug log data are useful for determinating protocol violations and can be used to determine treatment compliance. The drug log data are categorized as an intervention from a CDISC perspective.

Laboratory Data

Laboratory data may consist of many different collections of tests, such as ECG laboratory tests, microbiologic laboratory tests, and other therapeutic-indication-specific clinical lab tests. However, laboratory data traditionally consist of results from urinalysis, hematology, and blood chemistry tests. Traditional laboratory data can come from what are called local laboratories, which are labs at the clinical site, or from central laboratories where the clinical sites send their samples for analysis. Often when the laboratory data come from a central laboratory, there is no physical CRF page for the data and they are loaded into the clinical data management system directly from an electronic file. Local laboratory data may be represented with a CRF page such as this:

Protocol Name   Patient Number: _ _ _ - _ _ _   Visit Identifier

Laboratory Data:

Hematology				
Laboratory Test	**Date**	**Time (24-hour clock)**	**Result**	**Units**
Platelets	_/_/____	_ : _	_____	_____
Hemaglobin	_/_/____	_ : _	_____	_____
Hematocrit	_/_/____	_ : _	_____	_____
etc.				

Chemistry				
Laboratory Test	**Date**	**Time (24-hour clock)**	**Result**	**Units**
Serum Creatinine	_/_/____	_ : _	_____	_____
Total Cholesterol	_/_/____	_ : _	_____	_____
Basophils	_/_/____	_ : _	_____	_____
etc.				

Laboratory data can pose a challenge to the statistical programmer in many ways. Simply obtaining the data can sometimes be difficult. Occasionally you have to work with a specialized local laboratory, and sometimes just getting the data to the statistics group in a usable format can be hard. For example, the local laboratory staff may have used Microsoft Excel for data entry, and when they entered the data they entered rows within the columnar data with inconsistent formats, making machine readability of the resulting data file difficult. Another common issue is found within the "units" variable shown above. If local labs were used, it is likely that the lab units will have to be converted to a common unit for each laboratory test. Finally, laboratory values often need to be flagged as outside the normal range or perhaps outside the "clinical concern"/"panic range," where the latter is just a more extreme version of the former. Sometimes, the local or central laboratory flags these records, but it is not uncommon for the statistical programmer to have to make these assignments as well.

Laboratory data are most often associated with safety analyses, but they may play a part in efficacy analyses as well, especially if the laboratory data are part of the clinical endpoint definition. From a CDISC perspective, laboratory data are a finding, as they are a planned assessment.

Adverse Event Data

In the FDA's "Guidance for Industry E6 Good Clinical Practice: Consolidated Guidance," an adverse event is defined as follows:

> Any untoward medical occurrence in a patient or clinical investigation subject administered a pharmaceutical product and that does not necessarily have a causal relationship with this treatment. An AE can therefore be any unfavorable and unintended sign (including an abnormal laboratory finding), symptom, or disease temporally associated with the use of a medicinal (investigational) product, whether or not related to the medicinal (investigational) product.

The adverse event form is fairly standard across clinical trials. The form consists of a list of events for which data are entered as free text and are later coded with a dictionary such as MedDRA and some associated event attribute variables. In just about any clinical trial, an adverse event form very similar to the following sample will be found.

Protocol Name   Patient Number: _ _ _ - _ _ _

Adverse Events:

	Adverse Event	Start Date	End Date	Check If Ongoing	Severity	Action Taken	Drug Relation	Check If Serious
1	_____	_ / _ / _	_ / _ / _	__	_ Mild _ Moderate _ Severe	_ None _ Discontinued _ Dose changed _ Hospitalization _ Additional medication given	_ Yes _ No	__
2	_____	_ / _ / _	_ / _ / _	__	_ Mild _ Moderate _ Severe	_ None _ Discontinued _ Dose changed _ Hospitalization _ Additional medication given	_ Yes _ No	__
3	_____	_ / _ / _	_ / _ / _	__	_ Mild _ Moderate _ Severe	_ None _ Discontinued _ Dose changed _ Hospitalization _ Additional medication given	_ Yes _ No	__
	etc.							

The adverse event form is a cornerstone of patient safety monitoring, and as such it contains very important data. There are several data issues for the statistical programmer to be concerned about here.

Treatment-Emergent Signs and Symptoms

In guidance document ICH E3, "Structure and Content of Clinical Study Reports," the FDA defines *treatment-emergent signs and symptoms (TESS)* as "events not seen at baseline and events that worsened even if present at baseline." As simple as that may sound, it can sometimes be quite difficult to program. The important data variables that come into play are dosing record dates and times, adverse event start and stop times, and adverse event severity. All of these data variables need to be completed accurately for TESS to be calculated properly.

Serious Adverse Event Reconciliation

Just as there is an adverse event form, there is usually a *serious adverse event* (*SAE*) form. Note here that "serious" as defined by the FDA is different from "severe" on the adverse event form. A patient can have a "severe" headache that may not be considered "serious." The ICH guideline (also in ICH E3) entitled "Clinical Safety Data Management: Definitions and Standards for Expedited Reporting" defines serious adverse events as follows:

> A serious adverse event (experience) or reaction is any untoward medical occurrence that at any dose: results in death, is life-threatening, requires inpatient hospitalization or prolongation of existing hospitalization, results in persistent or significant disability/incapacity, or is a congenital anomaly/birth defect.

Usually a separate CRF is used to capture serious adverse events, as those must be reported to the FDA within 24 hours. That often means that the serious adverse events CRF data and the regular trial CRF adverse events are collected in different data tables, if not entirely different software systems. Pharmaceutical companies often want to reconcile the two databases to ensure that all serious adverse events appear in the regular-trial CRF adverse events database and that any event in the serious adverse events database is flagged properly as serious in the regular CRF adverse events database.

The problem is that the regular-trial adverse events database and the serious adverse events database do not join well if at all programmatically. You can attempt to join or merge the two databases by event start date and coded term, and that will join many regular-trial adverse events to the serious events. However, this is far from foolproof, because of mismatches in adverse event start dates and because the adverse events may have been coded slightly differently in the two systems. The best way to link the serious adverse events and adverse events databases is to have the clinical data management system create a linking variable key for you. In lieu of that, the only way to reliably link the two data sources is manually.

Concomitant Medication Reconciliation

Notice that in the sample adverse event form presented earlier, "additional medication given" is a possible action taken in response to an adverse event. Often you want to know precisely which medication was taken, but because that information may not be collected on the adverse event form, there needs to be a linkage with the concomitant medications form. Once again, this is not something than can reliably be done programmatically unless the clinical data management system creates a linking variable key behind the adverse event and concomitant medications forms. Some data management systems do this, but most do not.

Laboratory Data Reconciliation

The adverse event for a patient may indicate a medical condition such as hypercholestimia, so there may be a request to ensure that there are elevated cholesterol laboratory data to verify such a claim. You can sometimes make this kind of verification with programming if you know precisely which lab tests are involved and what level indicates a probable adverse event.

In the end, because of the importance of the data, it is imperative that the entire adverse event form data are cleaned. Reconciling the adverse event data with other clinical data in the clinical data management system can be very difficult if the data management system does not provide variable keys for linking such data. Adverse event data fall into the safety area of statistical analyses and are considered an event from a CDISC perspective.

Endpoint/Event Assessment Data

Endpoint or event assessments capture what the clinical trial was designed to study. For example, if a clinical trial were studying an anti-epilepsy medication, then the event form would likely collect seizure information. The endpoint or event assessment form is designed to collect data after the investigational drug or device intervention so that these data can be statistically compared to data from the patient's state before the drug or device intervention. Endpoint or event collection pages vary widely because of the broad range of ways to measure clinical disease, but here is a simplified sample endpoint collection page:

Protocol Name	Patient Number: _ _ _ - _ _ _	Visit Identifier

Endpoint Assessment:

Visit Date: _ _ / _ _ _ / _ _ _ _ (Day/Month/Year)
Did the patient have an event of interest? ☐ Yes   ☐ No
  If yes, what day did the event occur on? _ _ / _ _ _ / _ _ _ _ (Day/Month/Year)

In this form, "event" would be replaced by some clinical finding such as "myocardial infarction," "stroke," "seizure," or the like. This example form is extremely simplified, as there are usually a number of associated data variables captured as well. The event/endpoint page data must be clean, because it likely captures the primary efficacy data for the clinical trial.

The problem with endpoint data usually occurs when they need to be reconciled against data collected by the *clinical endpoint committee (CEC)*, which we discuss next. The endpoint/event data are almost always used for efficacy analyses but may be used for safety analyses as well. From a CDISC perspective, the endpoint/assessment is often considered a finding, as it is a planned examination, but it could also be considered an unplanned event.

Clinical Endpoint Committee (CEC) Data

It is often the case that the endpoint/event form captures data that are not entirely objective because they contain some level of clinical judgment. For instance, when precisely is a cold cured, was an event truly a myocardial infarction, or did any given event truly occur? The clinical site investigator may decide using his or her clinical judgment that a given event occurred, but often it is necessary to have an independent assessment of that event by another physician. This independent review helps to make sure that events are reported in a consistent way across multiple clinical sites for a clinical trial. Usually what happens is that a condition on the regular case report form "triggers" the release of a CEC form to be sent to the CEC. The CEC then takes the CEC form and verifies whether or not an actual event occurred based on the data available in the patient's clinical records at the given site. A sample CEC form follows:

Protocol Name   Patient Number: _ _ _ - _ _ _   Visit Identifier

---

Endpoint Assessment:

Did the patient have the event of interest? ☐ Yes   ☐ No

   If yes, on what day did the event occur? _ _ / _ _ _ / _ _ _ _ (Day/Month/Year)

Other supportive data fields go here to verify that the event happened.

Reviewer signature: _____   _ _ / _ _ _ / _ _ _ _ (Day/Month/Year)

In this CEC form, "event" would be replaced by some clinical finding such as "myocardial infarction," "stroke," "seizure," or the like. Once again, this form is extremely simplified, and there are usually a number of associated data variables captured that help to support the existence of the event.

The biggest problem for the statistical programmer when using CEC data is reconciling these data against the regular CRF endpoint/event data. This can be a difficult task, especially when you consider that a patient may have more than one event on a given day. Fortunately, because the endpoint/event data are so critical to a clinical trial, the quality of the reconciliation from the CEC form to the CRF form is not often relegated to some form of fuzzy data join. Usually there will be a definitive linkage via a key mapping data set that links the CEC event data to the CRF event data. However, if that key data set does not exist, then the statistical programmer must prepare for some difficult programming. It is also worth noting that the data from the adverse event forms, laboratory forms, and other forms, as well as a specific "event" form, may in fact trigger clinical events. This may add to the complexity of the reconciliation programming.

The clinical endpoint committee data are almost always used for efficacy analyses, but they may also be used for safety analyses. From a CDISC perspective, the endpoint/assessment is considered a finding, as it is a planned examination.

Study Termination Data

The study termination form collects patient exit information from the clinical trial. Here is a sample study termination form:

Protocol Name   Patient Number: _ _ _ - _ _ _

Study Termination:

Did the patient complete the study?   Yes ☐   No ☐
If not, please indicate why: ____ Adverse event
                                  ____ Study medication unsatisfactory
                                  ____ Subject withdrew consent to participate
                                  ____ Protocol violation
                                  ____ Death   __ / ___ / ____ (Day/Month/Year)
                                  ____ Lost to follow-up

Last day of study medication: __ / ___ / ____ (Day/Month/Year)
Investigator signature: __ / ___ / ____ (Day/Month/Year)

The study termination form data may be used for efficacy or safety analysis purposes. With regard to safety, if patients discontinue a study medication earlier than patients on standard therapy or placebo, then that is important to know. For efficacy analyses, patients who withdraw due to a lack of efficacy or adverse event may be precluded from being considered a treatment responder or success. Also, often the study termination date is used as a censor date in time-to-event analyses for therapy efficacy. Study termination forms play a key role in patient disposition summaries found at the start of a clinical study report. From a CDISC perspective, the study termination form is a finding.

Treatment Randomization Data

The *randomization* of a patient in a given therapy is the cornerstone of a randomized clinical trial. You may find these data in more than one place. They are often found within some form of *Interactive Voice Response System* (*IVRS*), but they may also be found in an electronic file containing the treatment assignments or on the CRF itself. If randomization data are found on the CRF, they usually consist only of the date of randomization for treatment-blinded trials. IVRS data are often found outside the confines of the clinical data management system and usually consist of the following three types of data tables.

Randomization Scheme Data Set

The *randomization scheme* assigns a therapy randomly across a study population based on various stratification factors such as site, *blocking factor*, and perhaps subject demographics. There is no actual patient assignment information in this data table. Here is an example of a randomization scheme with a block size of four and a *treatment ratio* of 2:2:

Index	Site	Block	Treatment
1	101	1	Study Medication
2	101	1	Placebo
3	101	1	Study Medication
4	101	1	Placebo
5	101	2	Placebo
6	101	2	Placebo
7	101	2	Study Medication
8	101	2	Study Medication

Notice that treatment is randomly assigned within the given blocks and that there are two placebos and two study medications in each block. Also notice the "index" variable. The order of the randomization scheme is critical to the usefulness of the scheme, as that is the order in which patients are assigned treatment. If the order of the scheme is altered in any way, then the scheme is damaged.

Drug Kit List Data Set

The *drug kit list* is simply a list that shows which drug container/kit label goes with which study medication. It might look something like this:

Kit Number	Treatment
10000001	Study Medication
10000002	Study Medication
10000003	Study Medication
10000004	Study Medication
10000005	Study Medication

Drug Assignment Data Set

The *drug assignment data set* indicates which patient got which drug. It might look something like this:

Site	Subject	Treatment
101	0001	Study Medication
101	0002	Study Medication
101	0003	Placebo
101	0004	Placebo

Note that the drug assignment data may not exactly match the order in the randomization scheme, as different patients pass screening procedures and are eligible for randomization at different times. Sometimes there are errors in treatment assignment that lead to a discrepancy between the drug assignment and the randomization scheme.

Other data sets may be found within the IVRS system that prove useful to the statistical programmer as well. Often the IVRS collects several baseline patient characteristics that are used in the stratification of the randomization scheme and subsequent assignment of study therapy. Finally, the preceding examples show in detail what the treatment variable is, in the "treatment" column. It is more often the case that the treatment variable is coded, such as "A" or "B" or "C." It is of paramount importance that you know with absolute certainty how the treatment code can be properly interpreted.

The randomization data are used in both efficacy and safety analyses, as they are typically the key stratification variable for the trial. The randomization data allow you to answer the question of whether patients who are getting the study therapy fare better than the alternative. CDISC places treatment assignment information in the special demographics domain.

Quality-of-Life Data

Sometimes you may also see *quality-of-life (QOL) data* collected for your clinical trial. Quality-of-life data are collected to measure the overall physical and mental well-being of a patient. These data are usually collected with a multiple-question patient questionnaire and may be summed up into an aggregate patient score for analysis. Some commonly used quality-of-life questionnaires are the SF-36 and SF-12 Health Survey, but there are quite a few disease-specific QOL questionnaires available to clinical researchers.

C h a p t e r 3

Importing Data

In most cases, the data that you use for clinical trial analyses are found in some kind of computer file external to the SAS System. The data you need may be found in a permanent SAS data set, a relational database table found in Oracle or Microsoft SQL Server, a Microsoft Access or Excel file, a simple delimited ASCII text file, or even an XML file. In any case SAS provides a wide array of ways in which to import data files into SAS. We explore these tools and the advantages and disadvantages of each in this chapter.

Importing Relational Databases and Clinical Data Management Systems

Most clinical data management systems used for clinical trials today store their data in *relational database* software such as Oracle or Microsoft SQL Server. A relational database is composed of a set of rectangular data matrices called "tables" that relate or associate with one another by certain key fields. The language most often used to work with relational databases is *structured query language* (*SQL*). The SAS/ACCESS SQL Pass-Through Facility and the SAS/ACCESS LIBNAME engine are the two methods that SAS provides for extracting data from relational databases.

SAS/ACCESS SQL Pass-Through Facility

The SAS/ACCESS SQL Pass-Through Facility has long been one of the only ways of getting data out of a relational database. It is still a flexible means of obtaining relational database data, as it allows for using SAS SQL as a means of filtering or modifying data on the way into SAS. Let's look at the following simple example.

Program 3.1 Using the SQL Pass-Through Facility to Get Data from Oracle

```
proc sql;
   connect to oracle as oracle_tables
         (user = USERID  orapw = PASSWORD  path = "INSTANCE");

   create table AE as
      select * from connection to oracle_tables
      (select * from AE_ORACLE_TABLE );

   disconnect from oracle_tables;
quit;
```

This code simply extracts the data from the table named "AE_ORACLE_TABLE" within Oracle and places it in its entirety in a SAS work library data set called AE. The USER, ORAPW, and PATH parameters are specific to the Oracle database settings at a particular site, so you would need to consult an Oracle database administrator to get the proper values. The good thing about the SQL Pass-Through Facility is that it lets you create a more desirable SAS data set with some slight modifications. For example, look at the following minor changes to Program 3.1.

Program 3.2 Using the SQL Pass-Through Facility to Get Selected Data from Oracle

```
proc sql;
   connect to oracle as oracle_tables
      (user = USERID  orapw = PASSWORD  path ="INSTANCE");

   create table library.AE as
      select * from connection to oracle_tables
      (select subject, verbatim, ae_date, pt_text
         from AE_ORACLE_TABLE
         where query_clean="YES");

   disconnect from oracle_tables;
quit;
```

Notice how the highlighted changes allow for a permanent SAS data set to be created containing only the variables desired and only the records that have all data queries resolved by data management.

SAS/ACCESS LIBNAME Statement

Beginning with SAS 7, a new SAS/ACCESS LIBNAME statement interface was available for accessing data in relational databases. With this convenient SAS enhancement, accessing tables in relational databases is just a single LIBNAME statement away. For example, the previous example of the SQL Pass-Through Facility can be distilled to the following LIBNAME statement and associated DATA step.

Program 3.3 Using the SAS/ACCESS LIBNAME Statement to Get Data from Oracle

```
libname oratabs oracle user=USERNAME
      orapw = PASSWORD  path = "@INSTANCE"  schema = TRIALNAME;

data adverse;
   set oratabs.AE_ORACLE_TABLE;
      where query_clean = "YES";
      keep subject verbatim ae_date pt_text;
run;
```

In this program the "oratabs" libref allows all of the tables found in that Oracle data *instance* to be treated like SAS data sets. This is a simple and fast way of accessing relational databases, and it requires no knowledge of SQL to implement.

Although the preceding examples import Oracle data, SAS/ACCESS can be used to access quite a number of relational databases, including Oracle, Microsoft SQL Server, Sybase, DB2, and Informix. The database-specific details on how to set up these SAS/ACCESS connections can be found in the SAS/ACCESS product documentation.

Importing ASCII Text

On occasion you may find that you need to import data from ASCII text files for analysis. In this section, traditional rectangular ASCII text files are discussed. Although XML files are composed of ASCII text, they pose unique challenges that are discussed later in this chapter. Some examples of ASCII text data that you might need to import include the following:

- Laboratory normal ranges
- Randomization data from an interactive voice response system (IVRS)
- Specialized coding dictionaries
- Supportive trial data not in the clinical data management system

SAS provides many ways of importing ASCII text files. These methods include using PROC IMPORT and the Import Wizard, the SAS DATA step, and SAS Enterprise Guide.

PROC IMPORT and the Import Wizard

The SAS IMPORT procedure (PROC IMPORT) provides a quick way to read an ASCII text file into SAS. You can call PROC IMPORT by typing in the SAS code, or you can use the convenient SAS Import Wizard to build the PROC IMPORT code for you. Let's start by looking at using the SAS Import Wizard to import the following pipe-delimited (using the character "|") *laboratory normal range* reference file:

```
Lab Test|Units|Sex|Low Age|High Age|Low Normal|High Normal
CREATININE|mg/dL|F|0|18|0.7|1.1
CREATININE|mg/dL|F|19|200|0.4|0.9
CREATININE|mg/dL|M|0|18|0.7|1.3
CREATININE|mg/dL|M|19|200|0.5|1.2
HGB|g/dL|F|0|18|11.0|14.9
HGB|g/dL|F|19|200|12.0|15.9
```

```
HGB|g/dL|M|0|18|10.0|15.6
HGB|g/dL|M|19|200|13.3|19.0
HCT|%|F|0|18|32.0|44.5
HCT|%|F|19|200|34.0|47.6
HCT|%|M|0|18|31.0|47.7
HCT|%|M|19|200|39.5|55.5
NEUTROPHILS|%|F|0|18|45.0|74.2
NEUTROPHILS|%|F|19|200|39.3|74.0
NEUTROPHILS|%|M|0|18|42.2|77.3
NEUTROPHILS|%|M|19|200|40.3|75.5
PLATELETS|x10^9/L|F|0|18|204|370
PLATELETS|x10^9/L|F|19|200|167|410
PLATELETS|x10^9/L|M|0|18|175|355
PLATELETS|x10^9/L|M|19|200|153|420
ALKALINE PHOSPHATASE|U/L|F|0|18|43|112
ALKALINE PHOSPHATASE|U/L|F|19|200|42|137
ALKALINE PHOSPHATASE|U/L|M|0|18|50|169
ALKALINE PHOSPHATASE|U/L|M|19|200|44|130
```

You begin the process in the interactive SAS windowing environment by selecting "File" from the toolbar and then "Import Data…" from the drop-down menu. You will see a window like the following:

Notice how "Delimited File" has been selected as a data source. Click "Next" and use the browse feature to select the file to import. Notice the important "Options" button under the file selection field, which will open the SAS Import: Delimited File Options window. Click the "Options" button and you will see a window like the following:

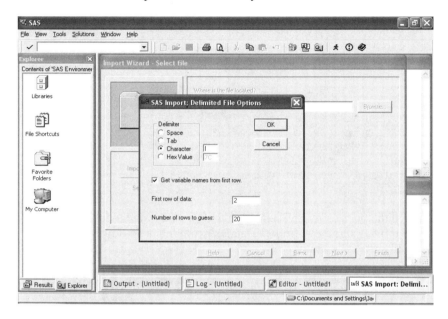

Notice that for this laboratory normal file you have to explicitly state that the delimiter is a pipe character. You also have to select "Get variable names from first row." Click "OK" and then "Next" and a window like the following will appear:

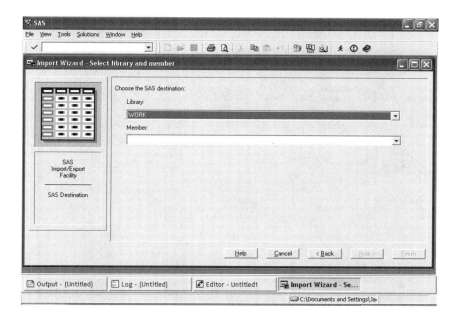

A SAS library can be selected from the drop-down list. For now, you will leave this as the WORK area and the member data set name will be set to "LABNORM." When you click "Next," a dialog box appears asking if you would like the PROC IMPORT procedure code to be saved. Saving is useful if you want to run the import process again without rerunning the Import Wizard.

Next, click "Finish" and the data will be imported into SAS.

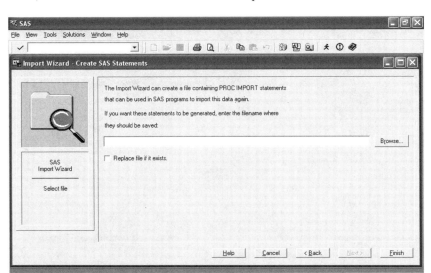

At this point the data have been imported as work.LABNORM. The data file looks like the following:

	Lab_Test	Units	Gender	Low_Age	High_Age	Low_Normal	High_Normal
1	CREATININE	mg/dL	F	0	18	0.7	1.1
2	CREATININE	mg/dL	F	19	200	0.4	0.9
3	CREATININE	mg/dL	M	0	18	0.7	1.3
4	CREATININE	mg/dL	M	19	200	0.5	1.2
5	HGB	g/dL	F	0	18	11	14.9
6	HGB	g/dL	F	19	200	12	15.9
7	HGB	g/dL	M	0	18	10	15.6
8	HGB	g/dL	M	19	200	13.3	19
9	HCT	%	F	0	18	32	44.5
10	HCT	%	F	19	200	34	47.6
11	HCT	%	M	0	18	31	47.7
12	HCT	%	M	19	200	39.5	55.5

(continued)

(continued)

13	NEUTROPHILS	%	F	0	18	45	74.2
14	NEUTROPHILS	%	F	19	200	39.3	74
15	NEUTROPHILS	%	M	0	18	42.2	77.3
16	NEUTROPHILS	%	M	19	200	40.3	75.5
17	PLATELETS	x10^9/L	F	0	18	204	370
18	PLATELETS	x10^9/L	F	19	200	167	410
19	PLATELETS	x10^9/L	M	0	18	175	355
20	PLATELETS	x10^9/L	M	19	200	153	420
21	ALKALINE PHOSP	U/L	F	0	18	43	112
22	ALKALINE PHOSP	U/L	F	19	200	42	137
23	ALKALINE PHOSP	U/L	M	0	18	50	169
24	ALKALINE PHOSP	U/L	M	19	200	44	130

Now look at rows 21–24. Notice how "ALKALINE PHOSPHATASE" is truncated to "ALKALINE PHOSP." This happens because the default behavior of the Import Wizard, PROC IMPORT, and the External File Interface (EFI) is that they scan only 20 rows deep into the file to determine variable attributes. Text field truncation is a common problem here. Another problem is that if a field appears to be numeric in the first 20 rows but later has character text beyond the scanning depth of PROC IMPORT, the procedure will terminate with an error message. There are two workarounds for this scanning depth problem.

The first workaround for the scanning depth problem is to simply increase the scanning depth. Beginning with SAS 8.2 the scanning depth of these tools can be increased globally via a registry edit according to SAS Note # SN-001075, which you can find at http://support.sas.com. A second approach that may be used in the preceding PROC IMPORT example is to increase the "Number of rows to guess" in the SAS Import: Delimited File Options window. A third approach is to increase the GUESSINGROWS option of the PROC IMPORT procedure beyond the default value of 20.

The second workaround for the scanning depth problem involves a bit of reverse engineering. When PROC IMPORT imports a delimited text file, it dynamically writes the SAS DATA step code to import the ASCII file into SAS. If SAS does not do exactly what you want during the import, you can always change the importing DATA step code that it creates. However, the DATA step code generated by SAS is not written as a SAS program. When you ask SAS to save the PROC IMPORT program statements that create the preceding file, you will see the following program.

Program 3.4 PROC IMPORT Code Written by the Import Wizard to Read an ASCII File

```
PROC IMPORT OUT= WORK.LABNORM
            DATAFILE= "C:normal_ranges.txt"
            DBMS=DLM REPLACE;
     DELIMITER='7C'x;
     GETNAMES=YES;
     DATAROW=2;
     GUESSINGROWS=20;
RUN;
```

What you want is the actual DATA step code that SAS writes behind the scenes to import the data. That code can be found in the Log window when the PROC IMPORT executes. For Program 3.4, the SAS log looks like this:

```
1     /**********************************************************
2     *    PRODUCT:    SAS
3     *    VERSION:    9.1
4     *    CREATOR:    External File Interface
5     *    DATE:       23FEB04
6     *    DESC:       Generated SAS Datastep Code
7     *    TEMPLATE SOURCE:  (None Specified.)
8     **********************************************************/
9          data WORK.LABNORM                                   ;
10         %let _EFIERR_ = 0; /* set the ERROR detection macro variable*/
11         infile 'C:normal_ranges.txt' delimiter = '|' MISSOVER DSD
11 ! lrecl=32767 firstobs=2 ;
12            informat Lab_Test $14. ;
13            informat Units $9. ;
14            informat Gender $1. ;
15            informat Low_Age best32. ;
16            informat High_Age best32. ;
17            informat Low_Normal best32. ;
18            informat High_Normal best32. ;
19            format Lab_Test $14. ;
20            format Units $9. ;
21            format Gender $1. ;
22            format Low_Age best12. ;
23            format High_Age best12. ;
```

(continued)

(continued)

```
24              format Low_Normal best12. ;
25              format High_Normal best12. ;
26       input
27                      Lab_Test $
28                      Units $
29                      Gender $
30                      Low_Age
31                      High_Age
32                      Low_Normal
33                      High_Normal
34       ;
35       if _ERROR_ then call symputx('_EFIERR_',1);   /* set ERROR
         detection macro variable */
36       run;

NOTE: The infile 'C:normal_ranges.txt' is:
      File Name=C:normal_ranges.txt,
      RECFM=V,LRECL=32767

NOTE: 24 records were read from the infile 'C:normal_ranges.txt'.
      The minimum record length was 22.
      The maximum record length was 40.
NOTE: The data set WORK.LABNORM has 24 observations and 7 variables.
NOTE: DATA statement used (Total process time):
      real time              0.24 seconds
      cpu time               0.01 seconds


24 rows created in WORK.LABNORM from C:normal_ranges.txt.
```

At this point, it is a simple exercise to copy the contents of this SAS log into the SAS Program Editor and make the changes that you want. In this case, at a minimum the INFORMAT value on "lab_test" should be increased from 14 to 20. If PROC IMPORT does not import the ASCII text file precisely as you want, it will get you 99% of the way to the desired result and leave you with a way to make the finishing touches by modifying the SAS code written to the SAS log.

SAS DATA Step

Before PROC IMPORT or the Import Wizard, the only tool available to import ASCII text files in SAS was the DATA step with INFILE and INPUT statements. This is not to say that the SAS DATA step is not a good tool for this task. In fact, the SAS DATA step is the most flexible and dynamic tool a statistical programmer has for importing a text file into SAS, but using it comes with a cost. That cost is usually the time spent troubleshooting the INFILE statement options and trying to remember the myriad tricks of importing text data. Nevertheless, here is what some SAS code might look like to import the LABNORM data in the previous example.

Program 3.5 Writing Custom SAS Code to Import Lab Normal Data

```
proc format;
   value $gender   "F" = "Female"
                   "M" = "Male";
run;

data labnorm;
   infile 'C:\normal_ranges.txt' delimiter = '|' dsd missover
          firstobs = 2;

   informat  Lab_Test $20.
             Units $9.
             Gender $1. ;

   format    Lab_Test $20.
             Units $9.
             Gender $gender.;

   input     Lab_Test $
             Units $
             Gender $
             Low_Age
             High_Age
             Low_Normal
             High_Normal;

   label     Lab_Test    = "Laboratory Test"
             Units       = "Lab Units"
             Gender      = "Gender"
             Low_Age     = "Lower Age Range"
             High_Age    = "Higher Age Range"
             Low_Normal  = "Low Normal Lab Value Range"
             High_Normal = "High Normal Lab Value Range";
run;
```

Note that there really is not much difference between the manual effort of using the custom SAS DATA step code and the effort of using PROC IMPORT. By writing a custom DATA step to import the data, I was able to add nice touches like detailed labels as well as formatting the gender variable with a descriptive permanent format. Since I manually determined the length of "lab_test," there was no field truncation issue. However, it would have saved time to start with PROC IMPORT and then just add the finishing touches to the SAS DATA step program that SAS generates.

SAS Enterprise Guide

Another approach to importing text files with SAS is to use the SAS Enterprise Guide interface, which ships free with Base SAS 9. The following example imports the LABNORM file using SAS Enterprise Guide 3.0.

From within a project in SAS Enterprise Guide, select "File" and then "Import Data..." from the toolbar. Select where to open the data from in the pop-up file selection window and click "Open." A data import options window like the following will appear:

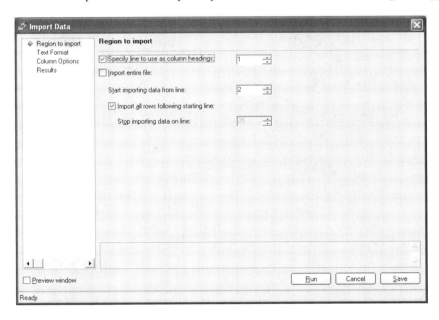

By default, "Import entire file" is selected, so deselect it and specify that line 1 is to be used for column headings and that all rows at line 2 and beyond are to be imported. Next, select the "Text Format" option and make the following changes so that the pipe character is set as the variable delimiter.

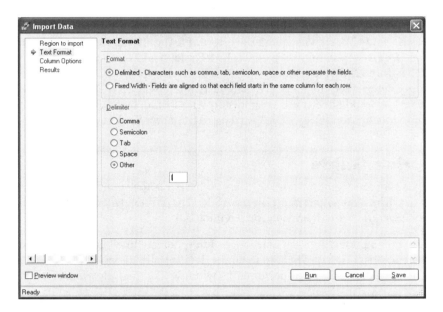

The next step is to specify "Column Options," where the SAS variable name, label, type, length, informat, and format can be changed if the SAS Enterprise Guide defaults are not desired. Also, there is a variable drop field option as well. The following is a sample of that window filled out for the LABNORM file.

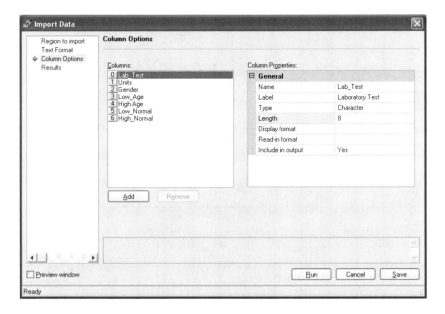

The Results window gives you a way to specify where the SAS data set will be stored. For this example, you modify it to look like the following to create a data set in SASWORK called LABNORM.

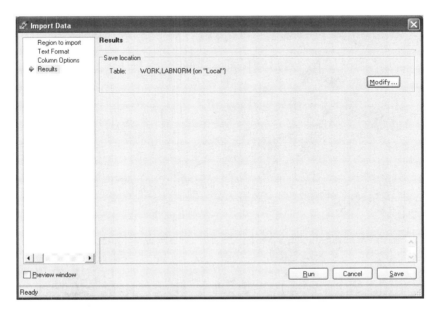

If you click "Save" in the window, the process is saved to the current project in SAS Enterprise Guide. If you select the "Preview window" in the lower-left corner of the window, you can see the SAS DATA step code that SAS Enterprise Guide generates and the results of that code. Click "Run" and SAS Enterprise Guide will import the data. If the results are not exactly what you want, then the saved "Import Data" process shown in the following window can be opened, changed, and rerun.

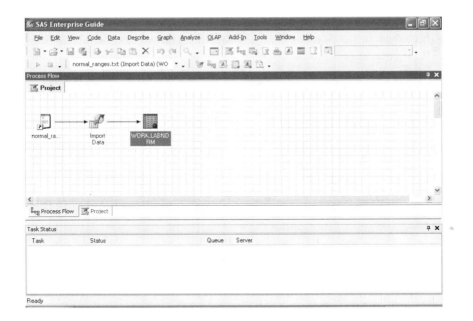

Importing Microsoft Office Files

Because Microsoft Office is so widely used, it is sometimes necessary for you to import data directly from Microsoft Excel or Microsoft Access. Excel files make for a poor "database," however. First, Excel files are almost guaranteed to come from a system that is not compliant with CFR 21 – Part 11. Second, it is often the case that the Excel files were created in such a way that the data are not *WYSIWYG* ("what you see is what you get"). In other words, each cell in Excel could be entered with a different Excel format, which you would not see until you either reformat an entire column in Excel or try to have some other software like SAS read the contents of the Excel file. For these reasons, it is best not to accept Microsoft Excel data as a data source for clinical trials if at all possible.

SAS provides several ways to read Microsoft Excel and Access files. We cover many of these import methods here using Microsoft formatted versions of the laboratory normal data used previously in this chapter. The examples here are based on the capabilities found in Base SAS and SAS/ACCESS for PC Files in SAS 9.1. In Microsoft Excel, the lab normal data file might look like the following:

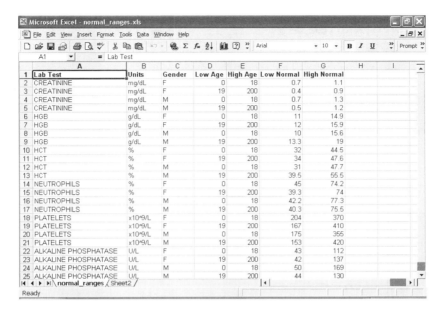

In Microsoft Access the lab normal data might look like this:

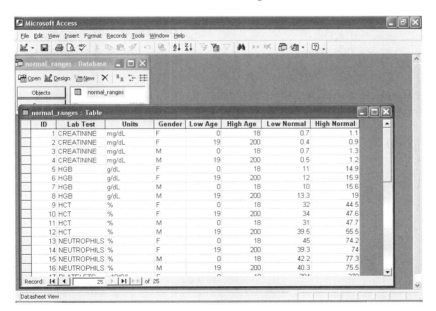

Some commonly used SAS tools for importing Microsoft Excel and Microsoft Access data into SAS include the LIBNAME statement, the Import Wizard/PROC IMPORT, the SQL Pass-Through Facility, and SAS Enterprise Guide.

LIBNAME Statement

Beginning with SAS 9.1, the LIBNAME statement can be used to simply map to an Excel or Access database. This facility is available in the Microsoft Windows and UNIX operating systems. For example, the following SAS code reads in and then prints the lab normal file normal_ranges.xls.

Program 3.6 Using the LIBNAME Statement to Read Microsoft Excel Data

```
libname xlsfile EXCEL "C:\normal_ranges.xls";
proc contents
   data = xlsfile._all_;
run;

proc print
   data = XLSFILE.'normal_ranges$'n;
run;
```

Note that the "EXCEL" engine specification is optional, because SAS would read the ".xls" extension in the physical filename and assume it indicates a Microsoft Excel file. Also note that the "xlsfile" libref refers to the entire Excel workbook. In the subsequent PROC PRINT, the "normal_ranges" must be specified so SAS will know which Excel worksheet to read. The data set/worksheet name in the PROC PRINT looks odd because of the existence of a special "$" character, which is normally not allowed as part of a data set name.

The normals_ranges.mdb Microsoft Access file could be read in with the following similar SAS code.

Program 3.7 Using the LIBNAME Statement to Read Microsoft Access Data

```
libname accfile ACCESS "C:\normal_ranges.mdb";

proc contents
   data = accfile._all_;
run;

proc print
   data = accfile.normal_ranges;
run;
```

Again, the "ACCESS" specification as a LIBNAME engine is optional, as the libref would default to Microsoft Access because ".mdb" is in the physical filename. Note that the ACCESS LIBNAME engine seems by default to import all text fields as 255 characters in length. Also note that all dates that come from Microsoft Access via the ACCESS engine are represented in SAS as SAS datetime fields. This is because Access has only datetime fields compared with SAS, which has date, time, and datetime variables.

SAS ACCESS and EXCEL librefs can be specified interactively by right-clicking on the Libraries icon in the SAS Explorer window and completing the parameters in the New Library window. You can define a libref called xlsfile that points to normal_ranges like this:

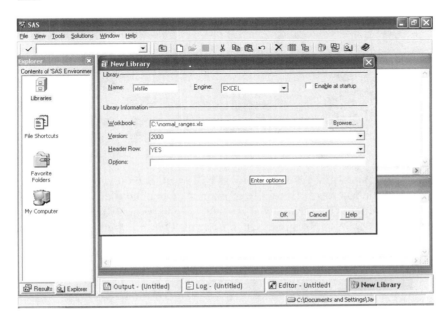

Be aware that the LIBNAME statement approach allows for both reading and writing to and from Microsoft Office files, which means the contents of the Microsoft Office files can be changed by SAS. There are many libref options not covered here that provide further access control features to Microsoft Excel and Access files. See the SAS documentation on the LIBNAME statement for PC files for more details.

Import Wizard and PROC IMPORT

The interactive SAS Import Wizard provides an easy way to import the contents of Microsoft Excel and Access files into SAS. Here again, the Import Wizard is essentially a graphical user interface that builds the PROC IMPORT code for you. Begin in the interactive SAS windowing environment by selecting "File" from the toolbar and then "Import Data…" from the drop-down menu. A window like the following will appear, where you can select Microsoft Excel as a standard data source.

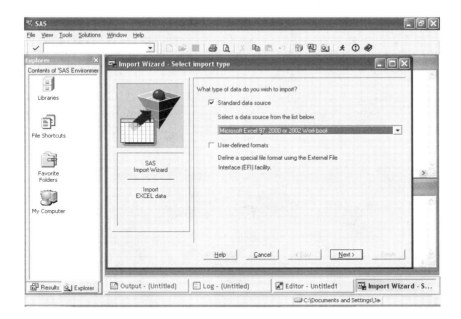

Click "Next," and a file browser window will open that allows for the drill-down and selection of the Microsoft Excel file of interest. Once the file is selected, a Select Table window will open. This window allows you to pick which worksheet in the Excel file you want to turn into a SAS data set. Click the "Options" button to see the new options available with SAS 9.1 and PROC IMPORT.

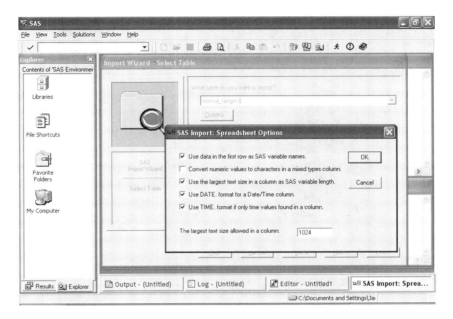

Note that the default options were chosen in the preceding window. We will look at those further when we explore the subsequent PROC IMPORT code. Now, click "OK" in the Spreadsheet Options window and then click "Next" in the Select Table window. The Select Library and Member window opens, which allows for the selection of a SAS library and data set name as follows.

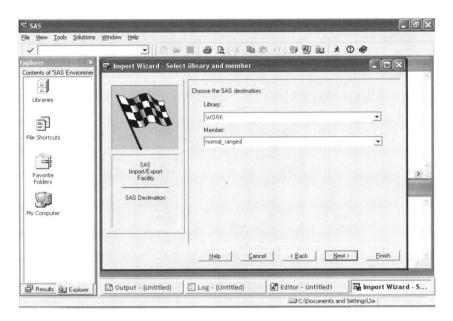

Click "Next" and SAS will prompt you to see if you want to save the PROC IMPORT code generated by the Import Wizard. Click "Finish" to complete the file import. Here is the PROC IMPORT code generated by SAS from this run.

Program 3.8 PROC IMPORT Code Generated by the Import Wizard to Read a Microsoft Excel File

```
PROC IMPORT OUT= WORK.normal_ranges
        DATAFILE= "C:\normal_ranges.xls"
        DBMS=EXCEL REPLACE;
   SHEET="normal_ranges$";
   GETNAMES=YES;
   MIXED=NO;
   SCANTEXT=YES;
   USEDATE=YES;
   SCANTIME=YES;
RUN;
```

If the options on PROC IMPORT do not produce what is desired, they can be changed and resubmitted, or the code can be saved to edit and run in batch mode later. Here are the new options available with SAS 9.1 along with my recommended settings.

Option	Purpose
DBSASLABEL	By default, the SAS label for an imported variable is set to the column name. Setting DBSASLABEL=NONE places null values into the SAS labels.
MIXED	Converts numeric values to character values if a column displays numeric and character text cells. Note that the default here is NO. Keep in mind that SAS scans only the first eight rows of the Excel column to determine whether the column is numeric or character. If SAS picks character and there are numeric cells later, then those will be set to blank. For this reason consider setting MIXED=YES.
SCANTEXT	When set to YES, this option tells SAS to scan the entire column to determine the width of the column. Always leave this set to YES.
SCANTIME	When set to YES, this option applies a SAS time format to a field if it appears to contain only time entries.
TEXTSIZE	This option hardcodes the maximum width of a character variable. It overrides SCANTEXT=YES.
USEDATE	When set to YES, this option formats SAS datetime fields with a date format. If you prefer to use datetime formats with datetime fields, set USEDATE=NO.

The Import Wizard process for Microsoft Access files works like the one for Excel files and produces similar PROC IMPORT code. Keep in mind that text fields get a default length of 255 characters when PROC IMPORT is used with Microsoft Access files. PROC IMPORT adds a number of file security options, as well as the ability to scan memo fields via the SCANMEMO option to determine the width of character fields in Microsoft Access files.

SAS/ACCESS SQL Pass-Through Facility

The SAS/ACCESS SQL Pass-Through Facility is another way for SAS to dynamically establish a connection to Microsoft Excel or Access files. You can connect to the Microsoft Excel file normal_ranges.xls by using the following SAS code.

Program 3.9 Using the SQL Pass-Through Facility to Read Microsoft Excel Data

```
**** OBTAIN AVAILABLE WORKSHEET NAMES FROM EXCEL FILE;
proc sql;
   connect to excel (path = "C:\normal_ranges.xls");
   select table_name from connection to excel(jet::tables);
quit;


**** GO GET NORMAL_RANGES WORKSHEET FROM EXCEL FILE;
proc sql;
   connect to EXCEL (path = "C:\normal_ranges.xls" header = yes
           mixed = yes  version = 2000 );
   create table normal_ranges as
      select * from connection to excel
      (select * from [normal_ranges$]);
   disconnect from excel;
quit;
```

Study the preceding SAS code. Notice how the first SQL step uses a special Microsoft Jet Engine query to obtain the names of the worksheet in normal_ranges.xls. Also note that the SQL step that fetches the normal_ranges worksheet from normal_ranges.xls does so by placing the worksheet in braces in the inner SELECT statement. The following SAS code uses the SQL Pass-Through Facility to connect to the Microsoft Access file normal_ranges.mdb.

Program 3.10 Using the SQL Pass-Through Facility to Read Microsoft Access Data

```
*** OBTAIN AVAILABLE TABLE NAMES FROM ACCESS FILE;
proc sql;
  connect to access (path = "C:\normal_ranges.mdb");
  select table_name from connection to access(jet::tables);
quit;
```

```
**** GO GET NORMAL_RANGES WORKSHEET FROM ACCESS FILE;
proc sql;
   connect to access (path="C:\normal_ranges.mdb");
   create table normal_ranges as
      select * from connection to access
      (select * from normal_ranges);
   disconnect from access;
quit;
```

Note that the SQL Pass-Through Facility to Microsoft Excel and Access files does default to 255 characters in length for character fields.

SAS Enterprise Guide

Another approach to importing Microsoft Excel and Access files with SAS is to use the SAS Enterprise Guide interface. The following example imports the normal_ranges.xls file using SAS Enterprise Guide 3.0.

From within a project in SAS Enterprise Guide, select "File" and then "Import Data…" from the toolbar. Select where to open the data from in the pop-up file selection window and click "Open." SAS Enterprise Guide knows by default that you want to open an Excel file and provides a worksheet selection window that looks like this:

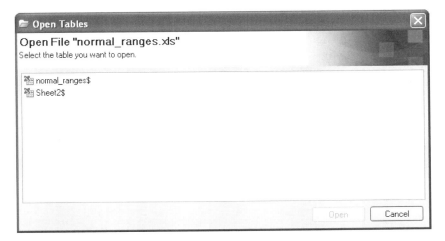

Next, simply select the normal_ranges$ sheet and click "Open" to advance to the Import Data window. That window looks like the following for your Excel file:

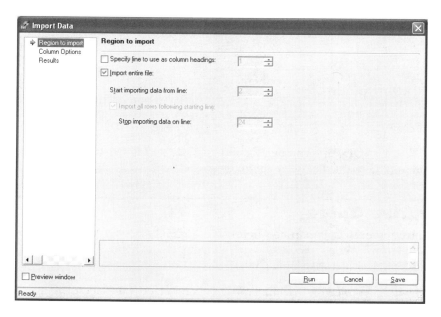

Note how SAS Enterprise Guide defaults to row 1 containing column headings and then begins importing actual data in column 2. If you select "Column Options" in the left pane, you will see this window:

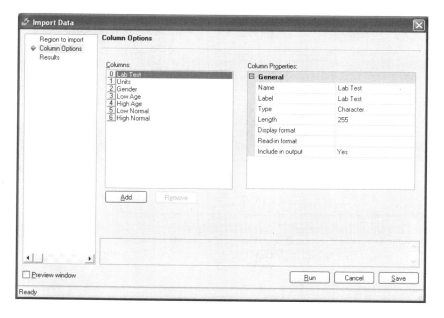

At this point, column/variable attributes can be changed. For instance, in the preceding window you see that the Lab Test field defaults to 255 characters wide, which you can easily reduce if desired. If you then click "Results" in the left pane, you can change where the SAS data set is stored. At this point, click "Run" and you will see that the import into SAS has taken place in SAS Enterprise Guide.

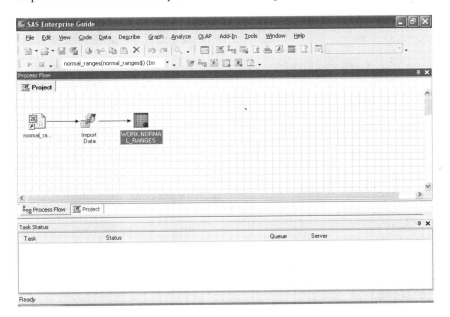

The process for importing Microsoft Access files into SAS Enterprise Guide is much like the process in the previous Excel example.

There are more SAS methods available for accessing PC files that include the DBF, DIF, ACCESS, and DBLOAD SAS procedures not covered in this text. Also, *Dynamic Data Exchange (DDE), Object Linking and Embedding (OLE),* and *Open Database Connectivity (ODBC)* methods can be used by SAS to exchange information in files as well as to control Microsoft Office applications from SAS. Note that some methods provide for one-way data importing into SAS, while others provide two-way read and write access to Microsoft Excel and Access files. Select the method that best suits your needs. For instance, if the source data is something that you do not want modified under any circumstances, then using PROC IMPORT might be better than setting up a dynamic two-way LIBNAME reference to a data file.

Importing XML

XML (E<u>x</u>tensible <u>M</u>arkup <u>L</u>anguage) is simply a way of "marking up," or labeling, ASCII text with special *markup tags* to give it structure. XML, a cousin of *HTML* and descendant of *SGML*, is a creation of the World Wide Web Consortium (W3C). XML continues to gain ground in many industries as a mechanism for describing and exchanging data, and for good reason. Here are just a few of the benefits of XML data:

- The data are stored in a non-proprietary open format, so no special software is needed to open it.

- The data are stored as simple ASCII text, so they are easily readable.

- Because the data are stored as simple ASCII text, text file browsers far into the future should still be able to read an XML file created today.

- XML can store non-rectangular-shaped data.

- Because XML is an open standard, many industries are developing open standards for XML data exchange. CDISC is the organization leading XML data standardization for the clinical trial industry.

Beginning with SAS 8, there are tools in SAS to help you read and write XML files. Starting with SAS 9.1, SAS provides XML import support via the XML LIBNAME engine (previously called SXLE) and the SAS XML mapper (previously called Atlas). We examine these tools for XML data import next. To begin, let's look at three records from the LABNORM file as they might appear in XML. Keep in mind that this data file could be represented in many different ways in XML, and this is just one representation.

```xml
<?xml version="1.0" ?>
<NORMALS>
  <ROW>
    <Lab_Test>CREATININE</Lab_Test>
    <Units>mg/dL</Units>
    <Sex>F</Sex>
    <Low_Age>0</Low_Age>
    <High_Age>18</High_Age>
    <Low_Normal>0.7</Low_Normal>
    <High_Normal>1.1</High_Normal>
  </ROW>
  <ROW>
    <Lab_Test>ALKALINE PHOSPHATASE</Lab_Test>
    <Units>mg/dL</Units>
    <Sex>F</Sex>
    <Low_Age>0</Low_Age>
    <High_Age>18</High_Age>
    <Low_Normal>43</Low_Normal>
```

```
      <High_Normal>112</High_Normal>
   </ROW>
   <ROW>
     <Lab_Test>NEUTROPHILS</Lab_Test>
     <Units>%</Units>
     <Sex>F</Sex>
     <Low_Age>0</Low_Age>
     <High_Age>18</High_Age>
     <Low_Normal>45.0</Low_Normal>
     <High_Normal>74.2</High_Normal>
   </ROW>
 </NORMALS>
```

There are two things that you may notice when looking at the lab normal data represented as XML. First, the file seems verbose. Whereas previously the lab normal file could be represented with three lines of pipe-delimited text, XML represents the same data with 30 lines of text. Second, you can read and somewhat understand the XML file just by looking at it if you know a markup language such as HTML or SGML. Let's look at how we can import these XML data into SAS.

XML LIBNAME Engine

The XML LIBNAME engine gives you the ability to read XML data using a simple LIBNAME statement. This sounds too good to be true, and perhaps it does state things a bit too simply. Because XML is extremely flexible in how it represents data, the SAS XML LIBNAME engine employs "XML map" files to help with data import. The XML map files are themselves valid XML files that describe the XML data file to be read. Here is an example of an XML map file that will help the XML LIBNAME engine to read the lab normal XML data.

```
<?xml version="1.0" encoding="windows-1252"?>
<XML_MAP version="1.2" name="XML_MAP">
  <TABLE name="NORMALS">
    <TABLE-PATH syntax="XPath">/NORMALS/ROW</TABLE-PATH>

    <COLUMN name="Lab_Test">
      <DESCRIPTION>Laboratory Test</DESCRIPTION>
      <PATH  syntax="XPath">/NORMALS/ROW/Lab_Test</PATH>
      <TYPE>character</TYPE>
      <DATATYPE>string</DATATYPE>
      <LENGTH>20</LENGTH>
    </COLUMN>

    <COLUMN name="Units">
      <DESCRIPTION>Lab Units</DESCRIPTION>
      <PATH  syntax="XPath">/NORMALS/ROW/Units</PATH>
```

```
        <TYPE>character</TYPE>
        <DATATYPE>string</DATATYPE>
        <LENGTH>5</LENGTH>
      </COLUMN>

      <COLUMN name="Sex">
        <DESCRIPTION>Sex</DESCRIPTION>
        <PATH  syntax="XPath">/NORMALS/ROW/Sex</PATH>
        <TYPE>character</TYPE>
        <DATATYPE>string</DATATYPE>
        <LENGTH>1</LENGTH>
      </COLUMN>

      <COLUMN name="Low_Age">
        <DESCRIPTION>Low Age Range</DESCRIPTION>
        <PATH  syntax="XPath">/NORMALS/ROW/Low_Age</PATH>
        <TYPE>numeric</TYPE>
        <DATATYPE>integer</DATATYPE>
      </COLUMN>

      <COLUMN name="High_Age">
        <DESCRIPTION>High Age Range</DESCRIPTION>
        <PATH  syntax="XPath">/NORMALS/ROW/High_Age</PATH>
        <TYPE>numeric</TYPE>
        <DATATYPE>integer</DATATYPE>
      </COLUMN>

      <COLUMN name="Low_Normal">
        <DESCRIPTION>Low Normal Lab Value Range</DESCRIPTION>
        <PATH  syntax="XPath">/NORMALS/ROW/Low_Normal</PATH>
        <TYPE>numeric</TYPE>
        <DATATYPE>double</DATATYPE>
      </COLUMN>

      <COLUMN name="High_Normal">
        <DESCRIPTION>High Normal Lab Value Range</DESCRIPTION>
        <PATH  syntax="XPath">/NORMALS/ROW/High_Normal</PATH>
        <TYPE>numeric</TYPE>
        <DATATYPE>double</DATATYPE>
      </COLUMN>
    </TABLE>
</XML_MAP>
```

Because the XML map file is valid XML itself, you can read how the SAS variables are translated from the XML lab normals data file. Once the XML map file is defined, you just need the simple SAS program that follows to read the lab normals XML file into SAS.

Program 3.11 Using the XML LIBNAME Engine to Read XML Data

```
filename  normals 'C:\normal_ranges.xml';
libname   normals xml xmlmap=XML_MAP;
filename  XML_MAP 'C:\xml_map.map';

proc contents
   data = normals.normals;
run;

proc print
   data = normals.normals;
run;
```

Program 3.11 yields the following SAS output:

```
                        The CONTENTS Procedure

        Data Set Name          NORMALS.NORMALS   Observations          .
        Member Type            DATA              Variables             7
        Engine                 XML               Indexes               0
        Created                .                 Observation Length    0
        Last Modified          .                 Deleted Observations  0
        Protection                               Compressed            NO
        Data Set Type                            Sorted                NO
        Label
        Data Representation    Default
        Encoding               Default

                  Alphabetic List of Variables and Attributes

 #  Variable     Type  Len  Format  Informat  Label

 5  High_Age     Num    8   F8.     F8.       High Age Range
 7  High_Normal  Num    8   BEST8.  BEST8.    High Normal Lab Value Range
 1  Lab_Test     Char  20   $20.    $20.      Laboratory Test
 4  Low_Age      Num    8   F8.     F8.       Low Age Range
 6  Low_Normal   Num    8   BEST8.  BEST8.    Low Normal Lab Value Range
 3  Sex          Char   1   $1.     $1.       Sex
 2  Units        Char   5   $5.     $5.       Lab Units

                                                        Low_      High_
 Obs  Lab_Test              Units  Sex  Low_Age High_Age Normal    Normal
   1  CREATININE            mg/dL   F       0       18     0.7       1.1
   2  ALKALINE PHOSPHATASE  mg/dL   F       0       18      43       112
   3  NEUTROPHILS           %       F       0       18      45      74.2
```

Notice how the XML engine is indicated in PROC CONTENTS and that the number of observations is set to missing. For further guidance refer to SAS Help and Documentation, where the XML LIBNAME engine is documented extensively.

SAS XML Mapper

Beginning with SAS 9.1, the SAS XML Mapper (previously known as Atlas) is a SAS stand-alone Java application supplied as part of Base SAS. The SAS XML Mapper is a graphical user interface assistant for building SAS XML Map files. Here is a display that shows the previous example's XML Map file being built from within the XML Mapper application:

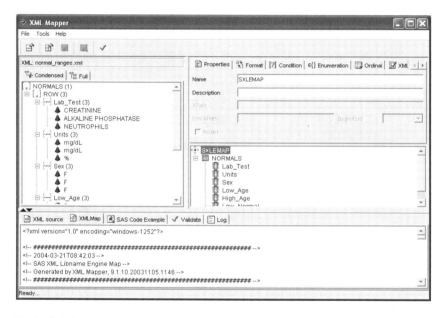

To build the XML map, table spaces (data sets) and columns (variables) can be dragged from the pane on the left and dropped in the pane on the right. There are several useful tabs available in the upper-right pane. They are as follows:

Properties tab	Allows for the definition of data sets and variables as well as the all-important observation boundary where the XML engine outputs data set observations.
Format tab	Lets you format variables further using SAS labels, formats, informats, and lengths.
Condition tab	Allows different attributes to be defined for the variable.
Enumeration tab	Establishes valid values for a given variable.

Ordinal tab	Establishes counting variable for the variable selected.
XMLMap Settings tab	Allows for XML file format encoding to be set and allows for reverse compatibility to previous versions of the XMLmap syntax.

At the bottom of the XML Mapper window are tabs that show the following: the text of the source XML file; the XML map as it is being derived; a SAS code template that can be used to import the XML data file; a validation tab showing the integrity of the XML map; and a log tab for the XML Mapper session. It is worth noting that the XML Mapper defaults to naming the created XML map as "SXLEMAP." This results in a warning when validating the XML Map that states, "Map name (SXLEMAP) is a default name. It is recommended that it be changed to avoid ambiguity." This is no cause for alarm, but you will probably want to change the XML map name from "SXLEMAP" to something more descriptive of your data file. Note that the warning message appears in the XML Mapper validation window as well as in the header of the resulting XML map file. The latter is shown in the following highlighted text:

```
<?xml version="1.0" encoding="windows-1252"?>

<!-- ##################################################### -->
<!-- 2004-03-21T08:57:57 -->
<!-- SAS XML Libname Engine Map -->
<!-- Generated by XML Mapper, 9.1.10.20031105.1146 -->
<!-- ##################################################### -->
<!-- ###   Validation report                        ### -->
<!-- ##################################################### -->
<!-- Map name (SXLEMAP) is a default name. It is recommended
that it be changed to avoid ambiguity. -->
<!-- Map validation completed - errors and/or warnings were
found. -->
<!-- ##################################################### -->

<SXLEMAP version="1.2" name="SXLEMAP">
```

XML will become more integral to the work of statistical programmers in the pharmaceutical industry as the standards, applications providers, and vendors make more use of this technology. Eventually you should expect the FDA to move away from SAS transport files to XML files as their standard data format for electronic data submission.

PROC CDISC

PROC CDISC is a new SAS procedure that is available as a hotfix for SAS 8.2 and ships as part of SAS 9.1.3. PROC CDISC is a procedure that allows you to import (and export) XML files that are compliant with the CDISC ODM version 1.2 schema. Here is a two-observation sample demographics ODM file that you might want to import into SAS:

```
<?xml version="1.0" encoding="windows-1252" ?>
<!--
     Clinical Data Interchange Standards Consortium (CDISC)
     Operational Data Model (ODM) for clinical data interchange

     You can learn more about CDISC standards efforts at
     http://www.cdisc.org/standards/index.html
  -->

<ODM xmlns="http://www.cdisc.org/ns/odm/v1.2"
     xmlns:ds="http://www.w3.org/2000/09/xmldsig#"
     xmlns:xsi="http://www.w3.org/2001/XMLSchema-instance"
     xsi:schemaLocation="http://www.cdisc.org/ns/odm/v1.2 ODM1-2-0.xsd"

     ODMVersion="1.2"
     FileOID="2004-09-11 Transfer of XT802"
     FileType="Snapshot"

     AsOfDateTime="2004-10-02T09:28:22"
     CreationDateTime="2004-10-02T09:28:22"
     SourceSystem="SAS 9.1"
     SourceSystemVersion="9.01.01M3P07282004">

  <Study OID="XT802">

     <!--
          GlobalVariables is a REQUIRED section in ODM markup
       -->
     <GlobalVariables>
        <StudyName>XT802</StudyName>
        <StudyDescription>Clinical Trial XT802 - Study of Infectious
                       Agent</StudyDescription>
        <ProtocolName>INVAG-XT802</ProtocolName>
     </GlobalVariables>

     <BasicDefinitions />

     <!--
          Internal ODM markup required metadata
       -->
     <MetaDataVersion OID="SDTMV3.1_01.00"
     Name="Submissions Data Tabulation Model Version 3.1 - Trial Meta Version
          01.00">
        <Protocol>
           <StudyEventRef StudyEventOID="BASELINE" OrderNumber="1"
            Mandatory="Yes" />
        </Protocol>
```

```
<StudyEventDef OID="BASELINE" Name="Study Event Definition"
                    Repeating="No"
 Type="Common">
   <FormRef FormOID="DM" OrderNumber="1" Mandatory="No" />
</StudyEventDef>

<FormDef OID="DM" Name="Form Definition" Repeating="No">
   <ItemGroupRef ItemGroupOID="DM" Mandatory="No" />
</FormDef>


<!--
     Columns defined in the table
  -->
<ItemGroupDef OID="DM" Repeating="No"
              SASDatasetName="DM"
              Name="Demographics"
              Domain="DM"
              Comment="Patient Demographics">
   <ItemRef ItemOID="ID.studyid" OrderNumber="1" Mandatory="No" />
   <ItemRef ItemOID="ID.usubjid" OrderNumber="2" Mandatory="No" />
   <ItemRef ItemOID="ID.siteid" OrderNumber="3" Mandatory="No" />
   <ItemRef ItemOID="ID.subjid" OrderNumber="4" Mandatory="No" />
   <ItemRef ItemOID="ID.domain" OrderNumber="5" Mandatory="No" />
   <ItemRef ItemOID="ID.rfstdtc" OrderNumber="6" Mandatory="No" />
   <ItemRef ItemOID="ID.brthdtc" OrderNumber="7" Mandatory="No" />
   <ItemRef ItemOID="ID.dmdtc" OrderNumber="8" Mandatory="No" />
   <ItemRef ItemOID="ID.age" OrderNumber="9" Mandatory="No" />
   <ItemRef ItemOID="ID.ageu" OrderNumber="10" Mandatory="No" />
   <ItemRef ItemOID="ID.dmdy" OrderNumber="11" Mandatory="No" />
   <ItemRef ItemOID="ID.country" OrderNumber="12" Mandatory="No" />
   <ItemRef ItemOID="ID.arm" OrderNumber="13" Mandatory="No" />
   <ItemRef ItemOID="ID.armcd" OrderNumber="14" Mandatory="No" />
   <ItemRef ItemOID="ID.sex" OrderNumber="15" Mandatory="No" />
   <ItemRef ItemOID="ID.rfendtc" OrderNumber="16" Mandatory="No" />
   <ItemRef ItemOID="ID.race" OrderNumber="17" Mandatory="No" />
</ItemGroupDef>
```

```
<!--
    Column attributes as defined in the table
  -->
<ItemDef OID="ID.studyid" SASFieldName="studyid" Name="Study Identifier"
 DataType="text" Length="20" />
<ItemDef OID="ID.usubjid" SASFieldName="usubjid"
 Name="Unique Subject Identifier" DataType="text" Length="40" />
<ItemDef OID="ID.siteid" SASFieldName="siteid" Name="Study Site Identifier"
 DataType="text" Length="10" />
<ItemDef OID="ID.subjid" SASFieldName="subjid"
 Name="Subject Identifier for the Study" DataType="text" Length="25" />
<ItemDef OID="ID.domain" SASFieldName="domain" Name="Domain Abbreviation"
 DataType="text" Length="2" />
<ItemDef OID="ID.rfstdtc" SASFieldName="rfstdtc"
 Name="Subject Reference Start Date/Time" DataType="text" Length="16" />
<ItemDef OID="ID.brthdtc" SASFieldName="brthdtc" Name="Date/Time of Birth"
 DataType="text" Length="16" />
<ItemDef OID="ID.dmdtc" SASFieldName="dmdtc" Name="Date/Time of Collection"
 DataType="text" Length="16" />
<ItemDef OID="ID.age" SASFieldName="age"
 Name="Age in AGEU at Reference Date/Time" DataType="float" />
<ItemDef OID="ID.ageu" SASFieldName="ageu" Name="Age Units"
 DataType="text" Length="6" />
<ItemDef OID="ID.dmdy" SASFieldName="dmdy" Name="Study Day of Collection"
 DataType="float" />
<ItemDef OID="ID.country" SASFieldName="country" Name="Country"
 DataType="text" Length="13" />
<ItemDef OID="ID.arm" SASFieldName="arm" Name="Description of Planned Arm"
 DataType="text" Length="16" />
<ItemDef OID="ID.armcd" SASFieldName="armcd" Name="Planned Arm Code"
 DataType="float" />
<ItemDef OID="ID.sex" SASFieldName="sex" Name="Sex" DataType="text"
 Length="1" />
<ItemDef OID="ID.rfendtc" SASFieldName="rfendtc"
 Name="Subject Reference End Date/Time" DataType="text" Length="16" />
<ItemDef OID="ID.race" SASFieldName="race" Name="Race" DataType="text"
 Length="9" />
    </MetaDataVersion>
  </Study>
```

```
<!--
     Administrative metadata
  -->
<AdminData />


<!--
     Clinical Data    : DM
                        Demographics
                        Patient Demographics
  -->
<ClinicalData StudyOID="TRIALXT802" MetaDataVersionOID="SDTMV3.1_01.00">
   <SubjectData SubjectKey="802101001">
      <StudyEventData StudyEventOID="BASELINE" StudyEventRepeatKey="1">
         <FormData FormOID="DM" FormRepeatKey="1">
            <ItemGroupData ItemGroupOID="DM" ItemGroupRepeatKey="1"
             TransactionType="Snapshot">
               <ItemData ItemOID="ID.studyid" Value="XT802" />
               <ItemData ItemOID="ID.usubjid" Value="802101001" />
               <ItemData ItemOID="ID.siteid" Value="101" />
               <ItemData ItemOID="ID.subjid" Value="802101001" />
               <ItemData ItemOID="ID.domain" Value="DM" />
               <ItemData ItemOID="ID.rfstdtc" Value="2003-01-10T12:52" />
               <ItemData ItemOID="ID.brthdtc" Value="1975-02-26" />
               <ItemData ItemOID="ID.dmdtc" Value="2003-01-10" />
               <ItemData ItemOID="ID.age" Value="27" />
               <ItemData ItemOID="ID.ageu" Value="YEARS" />
               <ItemData ItemOID="ID.dmdy" Value="1" />
               <ItemData ItemOID="ID.country" Value="United States" />
               <ItemData ItemOID="ID.arm" Value="Drug B" />
               <ItemData ItemOID="ID.armcd" Value="1" />
               <ItemData ItemOID="ID.sex" Value="M" />
               <ItemData ItemOID="ID.rfendtc" Value="2003-04-12T10:00" />
               <ItemData ItemOID="ID.race" Value="CAUCASIAN" />
            </ItemGroupData>
         </FormData>
      </StudyEventData>
   </SubjectData>
   <SubjectData SubjectKey="802101002">
      <StudyEventData StudyEventOID="BASELINE" StudyEventRepeatKey="1">
         <FormData FormOID="DM" FormRepeatKey="1">
            <ItemGroupData ItemGroupOID="DM" ItemGroupRepeatKey="1"
             TransactionType="Snapshot">
               <ItemData ItemOID="ID.studyid" Value="XT802" />
               <ItemData ItemOID="ID.usubjid" Value="802101002" />
               <ItemData ItemOID="ID.siteid" Value="101" />
               <ItemData ItemOID="ID.subjid" Value="802101002" />
               <ItemData ItemOID="ID.domain" Value="DM" />
               <ItemData ItemOID="ID.rfstdtc" Value="2003-01-10T18:40" />
               <ItemData ItemOID="ID.brthdtc" Value="1950-01-30" />
               <ItemData ItemOID="ID.dmdtc" Value="2003-01-10" />
               <ItemData ItemOID="ID.age" Value="52" />
               <ItemData ItemOID="ID.ageu" Value="YEARS" />
```

```
                     <ItemData ItemOID="ID.dmdy" Value="1" />
                     <ItemData ItemOID="ID.country" Value="United States" />
                     <ItemData ItemOID="ID.arm" Value="Drug A" />
                     <ItemData ItemOID="ID.armcd" Value="0" />
                     <ItemData ItemOID="ID.sex" Value="F" />
                     <ItemData ItemOID="ID.rfendtc" Value="2003-04-11T12:33" />
                     <ItemData ItemOID="ID.race" Value="CAUCASIAN" />
                  </ItemGroupData>
               </FormData>
            </StudyEventData>
         </SubjectData>
      </ClinicalData>
</ODM>
```

To import this ODM demographics file into SAS, you would run the following SAS code.

Program 3.12 Using PROC CDISC to Read ODM XML Data

```
        **** FILENAME POINTING TO ODM FILE;
        filename dmodm "C:\dm.xml";

        **** PROC CDISC TO IMPORT DM.XML TO DM WORK DATA SET;
        proc cdisc
            model               = odm
            read                = dmodm
            formatactive        = yes
            formatnoreplace     = no;

            odm
            odmversion          = "1.2"
            odmmaximumoidlength = 30
            odmminimumkeyset    = no;

            clinicaldata
            out                 = work.dm
            sasdatasetname      = "DM";
        run;
```

The resulting SAS work data set called "DM" looks like this:

Obs	__StudyOID	__MetaData VersionOID	__Subject Key	__StudyEvent OID	__StudyEvent RepeatKey
1	TRIALXT802	SDTMV3.1_01.00	802101001	BASELINE	1
2	TRIALXT802	SDTMV3.1_01.00	802101002	BASELINE	1

Obs	__FormOID	__FormRepeat Key	__ItemGroupOID	__ItemGroup RepeatKey	__TransactionType
1	DM	1	DM	1	Snapshot
2	DM	1	DM	1	Snapshot

Obs	studyid	usubjid	siteid	subjid	domain	rfstdtc	brthdtc	dmdtc
1	XT802	802101001	101	802101001	DM	2003-01-10T12:52	1975-02-26	2003-01-10
2	XT802	802101002	101	802101002	DM	2003-01-10T18:40	1950-01-30	2003-01-10

Obs	age	ageu	dmdy	country	arm	armcd	sex	rfendtc	race
1	27	YEARS	1	United States	Drug B	1	M	2003-04-12T10:00	CAUCASIAN
2	52	YEARS	1	United States	Drug A	0	F	2003-04-11T12:33	CAUCASIAN

Importing Files in Other Proprietary Data Formats

There was a time long ago, in the days of SAS 6 and earlier versions, when you needed to rely on software external to SAS to convert some data into SAS data sets so SAS could use it. Conceptual Software produced just such a tool with its DBMS/Copy package, which could easily convert many proprietary software database formats into SAS data sets. In 2002, SAS purchased the DBMS/Copy line of software tools and now distributes that software under its Dataflux subsidiary. So DBMS/Copy is still available for use, but the good news is that it is now largely unnecessary if you have SAS/ACCESS and SAS 9.

Base SAS 9 and SAS/ACCESS can read the following file formats directly into SAS:

Relational Databases	Non-Relational Databases
DB2	ASCII text including delimited files
Informix	ADABAS
OLE*	CA-DATACOM/DB
ODBC*	CA-IDMS
Oracle	IMS-DL/1
Sybase	SYSTEM 2000
MySQL	SAS/ACCESS interface to PC file formats reads
Microsoft SQL Server	Microsoft Access and Excel files directly
Teradata	

*.Can access many types of databases given an appropriate ODBC driver or OLE database.

However, there may be times when you need to read proprietary data formats into SAS that SAS 9 cannot handle directly. In that case, DBMS/Copy version 8 can convert the following list of proprietary data file formats:

4CaST/2
ABstat
Access
A-Cross
ACT! V1 Activity Files
ACT! V1 History Files
ACT! V2 Activities Files
ACT! V2 History Files
ACT! V2 Contacts Files
ACT! V2 Notes Files
ACT! V2 Expenses Files
Alpha IV Database
ASCII
Autobox
Autocast II
Axum
Bass
BMDP
BMDP 386
BMDP Dynamic
Business Cycle Indicators
Clarion
Clipper

CSS
CSS/3
DataEase (Forms Direct) 2.5
DataEase (Forms Direct) 4.0
DataEase (System/Form)
Datalex EntryPoint 90
dBASE II
dBASE III
dBASE IV
dBXL
DIF
Edwin
Egret
EpiInfo V3
EpiInfo V5
Excel 2
Excel 3
Excel 4
Excel 5 and 97
FOCUS
Forecast Plus
Forecast Pro
Forecast Pro (unv)

Forecast Pro (lotus)
FoxBASE+
Gauss
GLIM
GURU
HTML
Informix
Ingres
JMP
Knowledgeman
LIMDEP – Binary
LIMDEP - ASCII
Lotus 1-2-3
Lotus 1-2-3 V2
Lotus 1-2-3 V3
Lotus 1-2-3 V4
Matlab
MicroStat-II
Minitab
Multiplan SYLK Files
NCSS
Oracle (DOS Only)
Paradox 2 & 3

(continued)

Paradox 4+
PC-File
PFS:File
Probe/PC
PRODAS 3.2
PRODAS 4.0
Q&A 3.0
Quattro
Quattro Pro
Quattro Pro V5
Quattro Pro V6
Quattro Pro V7/8
Quicksilver
RATS
RATS V4
Rbase
Reflex V1
Reflex V2
SAS Xport Transport
SAS 5 Transport
SAS 6 Transport Compressed
SAS 6 Transport
SAS 8 Transport

SAS/PC
SAS 6 for Windows & OS/2
SAS 7, 8, 9 for Windows
SAS for UNIX
SAS 7, 8, 9 for UNIX
SCA
SigmaPlot V3.0
SigmaPlot V4.0
SigmaPlot V5.0
SigmaScan V2.5
SigmaScan V4.0
Smartware Database
SOLO
Soritec
S-Plus
SPSS/PC+
SPSS Portable
SPSS for OS/2
SPSS for Windows
Stata V2
Stata V2.1
Stata (8 byte doubles)
Stata V4/5 (4 byte doubles)

Stata V4/5 (8 byte doubles)
Stata V6 (4 byte doubles)
Stata V6 (8 byte doubles)
Stata V7 (4 byte doubles)
Stata V7 (8 byte doubles)
Stata Se (4 byte doubles)
Stata Se (8 byte doubles)
Statgraphics
Statgraphics for Windows
Statistica
Statistica V5
StatPac Gold
StatPlan III
Stats +
Statxact
SYGRAPH
Symphony V1
Symphony V1.1
SYSTAT
Visual Foxpro
X11
YStat

DBMS/Copy exists as a stand-alone program and also as a product called dfpower DBMS/Engines that allows conversion of these files from within SAS itself. If you have an ongoing need to pull a proprietary database format into SAS that SAS 9 with SAS/ACCESS alone cannot read, dfpower DBMS/Engines may be a solution.

C h a p t e r 4

Transforming Data and Creating Analysis Data Sets

Once the "raw" clinical data have been imported into SAS, the next step is to transform those raw data into more useful analysis-ready data. "Raw" data here mean data that have been imported without manipulation into SAS from another data source. That data source is likely to be a clinical data management system, but it could also be external laboratory data, IVRS data, data found in Microsoft Office files, or CDISC model data serving as the raw data. These raw data as they exist are often not ready for analysis. There may be additional variables that need to be defined, and the data may not be structured in a way that is required for a particular SAS analysis procedure. So once the raw data have been brought into SAS, they usually require some kind of transformation into analysis-ready files, which this chapter will discuss.

Key Concepts for Creating Analysis Data Sets

In this section we discuss some of the key concepts to keep in mind when creating an analysis data set. The next section takes these key concepts and puts them together to show how the most common analysis data sets are created.

Defining Variables Once

One of the primary reasons for creating analysis data sets is to have variable derivations in a single place. If a variable is defined in a single analysis data set, then the following are true:

- The variable is defined consistently.
- Any reviewer of the data can easily find the derivation.
- Any programmer can easily maintain the derivation.
- The derivation can be readily verified.

Some statistical programmers or statisticians may advocate placing new variable derivations within individual analysis or summary programs. I believe this is a practice that should be avoided. Imagine if you wanted to change the derivation of a single variable but you had to search 100 programs to find it and subsequently change it 100 times. Also, if the variable is defined 100 times, then the odds that it is defined the same way across 100 programs are low. Finally, if the derivation is stored once in a permanent analysis data set, then it can be verified easily. The same cannot be said of a variable that is derived in a summary program that disappears from memory when the SAS program terminates.

Defining Study Populations

Which patients should be in which data set is something that should be considered before analysis data sets are created. For example, it is often decided that all analysis data sets should have a record for a subject if that subject was randomized to treatment and is considered an intent-to-treat subject. Whether this is true or not, the specifications for analysis data sets should make it clear who should be present in any analysis data set. Here is a list of common populations and their definitions:

Population	Definition
Intent-to-treat	All patients randomized to a study therapy. It is intended that they will be treated. Patients are analyzed according to randomized treatment group.
Per-protocol	All patients who did not experience a subjectively defined serious protocol violation during the study.
Safety	All patients who actually received the study drug.
As-treated	Patients analyzed according to the study intervention they actually received. Patients may get a treatment that they were not randomized to.

Defining Baseline Observations

"Baseline" is a common clinical concept. You might hear your physician say, "Let's get a baseline cholesterol reading for you." The idea behind a baseline measurement is to determine the state of a patient before some expected event so that a subsequent comparison to that state can be made. For instance, you and your physician may want to get a baseline cholesterol reading early in life, as cholesterol typically increases with age. In clinical trials, you want to obtain a baseline measurement before a medical intervention to see what kind of effect the intervention had. Usually, the baseline value is the last reading prior to medical intervention. The following figure illustrates this concept for cholesterol measurements:

Cholesterol Measurement Date	Date of Initial Therapeutic Intervention
05JAN2000	
06JAN2000	
07JAN2000	09JAN2000
10JAN2000	
09FEB2000	

(Baseline Assessment → 07JAN2000)

Last Observation Carried Forward (LOCF)

Often you need to "carry forward" data to a specific time point due to holes or sparseness of data. The previous example on determining baseline cholesterol level provides an excellent context for this problem. Assume that you have several cholesterol readings of HDL, LDL, and triglycerides for patients before they take an experimental pill designed to reduce cholesterol levels. For each cholesterol parameter, you want the last observation carried forward so long as the measures occur within a five-day window before the pill is taken. Here are some sample data that illustrate the problem:

Subject	Sample Date	HDL	LDL	Triglycerides	Dosing Date
101	05SEP2003	48	188	**108**	
101	06SEP2003	**49**	**185**	.	
101					07SEP2003
102	01OCT2003	54	200	350	
102	02OCT2003	**52**	.	**360**	
103					07OCT2003

(continued)

(continued)

Subject	Sample Date	HDL	LDL	Triglycerides	Dosing Date
103	10NOV2003	.	**240**	900	
103	11NOV2003	30	.	**880**	
103	12NOV2003	**32**	.	.	
103	13NOV2003	35	289	930	13NOV2003

The underlined values in the sample data fit the definition of baseline. You can see how the last non-missing value should be carried forward. The following SAS code selects those proper baseline values.

Program 4.1 Deriving Last Observation Carried Forward (LOCF) Variables

```
**** INPUT SAMPLE CHOLESTEROL DATA.
**** SUBJECT = PATIENT NUMBER, SAMPDATE = LAB SAMPLE DATE,
**** HDL = HDL, LDL = LDL, AND TRIG = TRIGLYCERIDES.;
data chol;
input subject $ sampdate date9. hdl ldl trig;
datalines;
101 05SEP2003 48 188 108
101 06SEP2003 49 185 .
102 01OCT2003 54 200 350
102 02OCT2003 52 .   360
103 10NOV2003 .   240 900
103 11NOV2003 30 .   880
103 12NOV2003 32 .   .
103 13NOV2003 35 289 930
;
run;


**** INPUT SAMPLE PILL DOSING DATA.
**** SUBJECT = PATIENT NUMBER, DOSEDATE = DRUG DOSING DATE.;
data dosing;
input subject $ dosedate date9.;
datalines;
101 07SEP2003
102 07OCT2003
103 13NOV2003
;
run;
```

```
**** SORT CHOLESTEROL DATA FOR MERGING WITH DOSING DATA.;    ❶
proc sort
   data = chol;
      by subject sampdate;
run;

**** SORT DOSING DATA FOR MERGING WITH CHOLESTEROL DATA.;
proc sort
   data = dosing;
      by subject;
run;

**** DEFINE BASELINE HDL, LDL, AND TRIG VARIABLES;           ❷
data baseline;
   merge chol dosing;
      by subject;

      keep subject b_hdl b_ldl b_trig;

      **** SET UP ARRAYS FOR BASELINE VARIABLES AND LAB VALUES;
      array base {3} b_hdl b_ldl b_trig;
      array chol {3}   hdl   ldl   trig;

      **** RETAIN NEW BASELINE VARIABLES SO THEY ARE PRESENT
      **** AT LAST.SUBJECT BELOW.;
      retain b_hdl b_ldl b_trig;

      **** INITIALIZE BASELINE VARIABLES TO MISSING.;
      if first.subject then
         do i = 1 to 3;
            base{i} = .;
         end;

      **** IF LAB VALUE IS WITHIN 5 DAYS OF DOSING, RETAIN IT AS
      **** A VALID BASELINE VALUE.;
      if 1 <= (dosedate - sampdate) <= 5 then                 ❸
         do i = 1 to 3;
            if chol{i} ne . then
               base{i} = chol{i};
         end;
```

```
**** KEEP LAST RECORD PER PATIENT HOLDING THE LOCF VALUES.;
if last.subject;

label b_hdl  = "Baseline HDL"
      b_ldl  = "Baseline LDL"
      b_trig = "Baseline triglycerides";
run;
```

Notes for the program:

❶ The cholesterol data set is also sorted by sample date. This is critical, as the samples must be in the proper chronological order for the DATA step that follows.

❷ This DATA step uses ARRAYs, RETAINs of the newly created baseline values, and a final subsetting IF to keep the proper baseline cholesterol variables.

❸ If the cholesterol measurement is non-missing and was taken within the five days prior to drug dosing, then the cholesterol values are valid values for baseline. Note that because the cholesterol data are sorted chronologically (as mentioned in note 1), the last non-missing value within the five-day window is carried forward in time as the baseline value.

This is the resulting "baseline" data set:

subject	b_hdl	b_ldl	b_trig
101	49	185	108
102	52	.	360
103	32	240	880

Defining Study Day

Time is a critical measure for clinical trial analysis. Time is captured in clinical trial databases in a "*study day*" variable. Study day can be defined as the number of days from therapeutic intervention to any given time point or event. By defining study day, you create a common metric for measuring time across a population of patients in a clinical trial. There can be a study day calculation for any time point of interest. Adverse event start, study termination, and clinical endpoint event date all make good choices for study day calculations. The study day calculation is performed with one of the two following approaches.

Program 4.2 Calculating a Continuous Study Day

```
study_day = event_date - intervention_date + 1;
```

Here is the study_day result from Program 4.2:

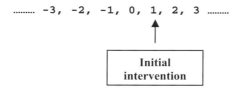

Program 4.3 Calculating a Study Day without Day Zero

```
if event_date < intervention_date then
    study_day = event_date - intervention_date;
else if event_date >= intervention_date then
    study_day = event_date - intervention_date + 1;
```

Here is the study_day result from Program 4.3:

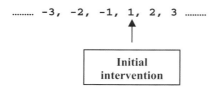

In Program 4.2, study_day is a continuous stream of integers with the date of initial intervention representing study day 1. Program 4.2 has the benefit of being a contiguous string of numbers that can be useful when graphing or calculating durations that span the day before the therapeutic intervention day. Although Program 4.3 appears more complex than Program 4.2, the only difference is that the day before initial intervention is study day –1 and not 0. Study day 0 does not exist in Program 4.3. Because of this, Program 4.3 may seem a bit more intuitive because the day before intervention is represented by study day "–1."

Both approaches to study day calculation are used in the pharmaceutical industry, although the algorithm in Program 4.3 is probably used more often. The CDISC Submission Data Tabulation Model states that you should use the algorithm in

Program 4.3 for creating study day variables for the SDTM data sets. However, the "General Considerations" document from the CDISC Analysis Data Set Modeling Team states that you should use the algorithm in Program 4.2 for analysis data sets. Whether you are deriving data based on the CDISC models or not, you should calculate study day variables in a consistent fashion across a clinical trial or set of trials for an application.

Windowing Data

Windowing is the process of taking an observation and placing a tag on it based on a range of acceptable values. The range of acceptable values could be dates or some other numerical measure. A tag is some descriptive label such as "Visit 5," "Baseline," or "Abnormal." Because dates are the data element most often associated with windowing practices, we look at an example of date windowing for visit assignment here.

Imagine that you are receiving laboratory data from a central laboratory and that in the data there is no reliable visit assignment. This is a common laboratory data problem. You need a valid visit assignment because you want to analyze the data by study visit. You know when patients received initial study medication, and you know that lab data should have been drawn at target times of baseline, 30 days, 60 days, 1 year, and 2 years following initial treatment. You want to capture the observations closest to the target dates, with earlier dates taking preference over later dates in the case of ties. Baseline observations must occur before initial drug dosing. With this information, a windowing algorithm might look like the following.

Program 4.4 Deriving a Visit Based on Visit Windowing

```
**** INPUT SAMPLE LAB DATA.
**** SUBJECT = PATIENT NUMBER, LAB_TEST = LABORATORY TEST NAME,
**** LAB_DATE = LAB COLLECTION DATE, LAB_RESULT = LAB VALUE.;
data labs;
input subject $ lab_test $ lab_date lab_result;
datalines;
101 HGB 999 1.0
101 HGB 1000 1.1
101 HGB 1011 1.2
101 HGB 1029 1.3
101 HGB 1030 1.4
101 HGB 1031 1.5
101 HGB 1058 1.6
101 HGB 1064 1.7
101 HGB 1725 1.8
101 HGB 1735 1.9
;
run;
```

```
**** INPUT SAMPLE DOSING DATE.
**** SUBJECT = PATIENT NUMBER, DOSE_DATE = DATE OF DOSING.;
data dosing;
input subject $ dose_date;
datalines;
101 1001
;
run;

**** SORT LAB DATA FOR MERGE WITH DOSING;
proc sort
   data = labs;
      by subject;
run;

**** SORT DOSING DATA FOR MERGE WITH LABS.;
proc sort
   data = dosing;
      by subject;
run;

**** MERGE LAB DATA WITH DOSING DATE. CALCULATE STUDY DAY AND
**** DEFINE VISIT WINDOWS BASED ON STUDY DAY.;
data labs;
   merge labs(in = inlab)
         dosing(keep = subject dose_date);
      by subject;

      **** KEEP RECORD IF IN LAB AND RESULT IS NOT MISSING.;
      if inlab and lab_result ne .;

      **** CALCULATE STUDY DAY.;
      if lab_date < dose_date then
         study_day = lab_date - dose_date;
      else if lab_date >= dose_date then
         study_day = lab_date - dose_date + 1;

      **** SET VISIT WINDOWS AND TARGET DAY AS THE MIDDLE OF THE
      **** WINDOW.;
      if . < study_day < 0 then                              ❶
         target = 0;
      else if 25 <= study_day <= 35 then
         target = 30;
      else if 55 <= study_day <= 65 then
         target = 60;
      else if 350 <= study_day <= 380 then
         target = 365;
```

```
            else if 715 <= study_day <= 745 then
               target = 730;

            **** CALCULATE OBSERVATION DISTANCE FROM TARGET AND ❷
            **** ABSOLUTE VALUE OF THAT DIFFERENCE.;
            difference = study_day - target;
            absdifference = abs(difference);
   run;

   **** SORT DATA BY DECREASING ABSOLUTE DIFFERENCE AND ACTUAL
   **** DIFFERENCE WITHIN A VISIT WINDOW.;
   proc sort
      data=labs;
         by subject lab_test target absdifference difference;
   run;

   **** SELECT THE RECORD CLOSEST TO THE TARGET AS THE VISIT.
   **** CHOOSE THE EARLIER OF THE TWO OBSERVATIONS IN THE EVENT OF
   **** A TIE ON BOTH SIDES OF THE TARGET.;
   data labs;
      set labs;
         by subject lab_test target absdifference difference; ❸

         if first.target and target ne . then
            visit_number = target;
   run;
```

Notes for the code:

❶ The visit "windows" are defined here. The definitions for valid windows are somewhat arbitrary and are typically based on clinical and statistical judgment from trial to trial. Here, the baseline window is defined as the set of observations before initial dosing. The one- and two-month windows are 30 and 60 days from dosing, respectively, plus or minus five days. Finally, the one- and two-year visits are defined as 365 and 730 days, respectively, plus or minus 15 days. Also note that a "target" variable is defined that is the center of the visit window and the best study day to get the observation.

❷ Here you calculate the "difference" variable, which is the distance from the lab observation to the "target" or most desirable place for the lab observation. You also create an "absdifference" variable, which is the absolute distance from the lab observation to the desired target.

❸ In this final step, if the record falls within a desired visit window (target is not missing), you take the first record within the window as the observation you want to use for analysis for that visit.

The results of the preceding SAS code would yield a data set called "labs" that would look like this after sorting on "subject" and "lab_date":

subject	lab_test	lab_date	Lab_ result	dose _date	days	target	difference	abs difference	visit_ number
101	HGB	999	1.0	1001	-2	0	-2	2	.
101	HGB	1000	1.1	1001	-1	0	-1	1	0
101	HGB	1011	1.2	1001	11	.	.	.	.
101	HGB	1029	1.3	1001	29	30	-1	1	.
101	HGB	1030	1.4	1001	30	30	0	0	30
101	HGB	1031	1.5	1001	31	30	1	1	.
101	HGB	1058	1.6	1001	58	60	-2	2	60
101	HGB	1064	1.7	1001	64	60	4	4	.
101	HGB	1725	1.8	1001	725	730	-5	5	730
101	HGB	1735	1.9	1001	735	730	5	5	.

To make this example easier to view, "visit_number" is set to "target," "lab_date" is not formatted as a SAS date, and "lab_result" is not realistic lab data but a usable row counter here. Typically "visit_number" would be an integer incremented by one for each subsequent visit.

Although this example is a common windowing algorithm, there are many others. For instance, you might exclude any observations that happened after the target study day. You might take the average of the values within any given window as the imputed result to analyze. No matter what windowing approach you take, the algorithms used to window data should be well documented so that a reviewer can understand what was done.

Transposing Data

Data transposition is the process of changing the orientation of the data from a normalized structure to a non-normalized structure or vice versa. There are many definitions of normalization of data, and you should learn about normal forms and normalization. Here, in brief, normalization of data means the process of taking information out of the variable definitions and turning that information into row definitions/keys in order to reduce the overall number of variables. Normalized data may also be described as "stacked," "vertical," or "tall and skinny," while non-normalized data are often called "flat," "wide," or "short and fat."

Graphically, the process of data transposition from normalized to non-normalized data looks like this:

Normalized Form   **Non-normalized Form**

Test	Result
Test1	1.0
Test2	2.0
Test3	3.0
Test4	4.0
Test5	5.0
Test6	6.0

Transposed

Test1	Test2	Test3	Test4	Test5	Test6
1.0	2.0	3.0	4.0	5.0	6.0

Typically, clinical data come to you in a shape that is dictated by the underlying CRF design and the clinical data management system. Most clinical data management systems use a relational data structure that is normalized and optimized for data management. Much of the time these normalized data are in a structure that is perfectly acceptable for analysis in SAS. However, sometimes the data need to be denormalized for proper analysis in SAS.

A problem occurs when end users of the data cannot conceptualize how to handle normalized data. These users go out of their way to denormalize any normalized data that they see. I have seen entire databases denormalized so that a user could work with the data, and in some cases the user unknowingly renormalizes the data so that he or she can then analyze it properly. This type of user needs to be coached as to when denormalization is needed.

Denormalization of data is needed when a statistical procedure requires that the information to be analyzed must be on the same observation. Procedures in SAS that perform data modeling are often the ones that require denormalized data, as they require that the *dependent variable* be present on the same observation as the *independent variables*. For example, imagine that you are trying to determine a mathematical model that predicts under what conditions a therapy is successful. That model might look like this:

$$\text{success} = \beta_0 + \beta_1(\text{age}) + \beta_2(\text{clinical therapy}) + \beta_3(\text{ethnicity}) + \beta_4(\text{systolic blood pressure at baseline}) + \beta_5(\text{systolic blood pressure at 6 months})$$

Here we see that we have a "success" variable that is dependent on a series of other variables. All of those data need to be present on a given observation in order for a statistical modeling procedure such as PROC LOGISTIC to be useful. The denormalized data set might look something like the following:

Subject ID	Success	Age	Clinical_ therapy	Ethnicity	Baseline SBP	6 month SBP
101001	1	34	1	2	75	78
101002	0	72	0	3	70	80

Do not denormalize data unless required by statistical procedures or if SAS BY processing of the data will meet your needs. Sometimes you can see just by looking at the reporting desired that denormalization is not required. For example, look at the following table requirement:

Summary of Clinical Success			
Visit	**Success**	**Active Therapy**	**Placebo**
Week 1	Success	N (%)	N (%)
	Failure	N (%)	N (%)
Week 2	Success	N (%)	N (%)
	Failure	N (%)	N (%)
Month 1	Success	N (%)	N (%)
	Failure	N (%)	N (%)
Month 2	Success	N (%)	N (%)
	Failure	N (%)	N (%)

In this table you can see that clinical success is summarized for each treatment by visit. The key here is "by visit." If the data set to be summarized is simply sorted by visit, then PROC FREQ, PROC TABULATE, or some other procedure can be executed with a BY VISIT statement. If the data set were denormalized, then the task of producing the required summary would be more difficult.

There are two tools commonly used for performing data transpositions in SAS: PROC TRANSPOSE, and a DATA step with ARRAY statements. PROC TRANSPOSE is a powerful tool that "flips" a data set with just a few lines of SAS code. On the other hand, DATA steps that employ arrays for data transposition are more flexible, in that they allow for more precise control of the transposition process. We will now examine PROC

TRANSPOSE to transform a SAS data set of systolic blood pressures. Let's assume that you have a normalized file and you want to end up with a single record per subject, with five variables holding the systolic blood pressure value for each visit. Using PROC TRANSPOSE, your program would look like the following.

Program 4.5 Transposing Data with PROC TRANSPOSE

```
**** INPUT SAMPLE NORMALIZED SYSTOLIC BLOOD PRESSURE VALUES.
**** SUBJECT = PATIENT NUMBER, VISIT = VISIT NUMBER,
**** SBP = SYSTOLIC BLOOD PRESSURE.;
data sbp;
input subject $ visit sbp;
datalines;
101 1 160
101 3 140
101 4 130
101 5 120
202 1 141
202 3 161
202 4 171
202 5 181
;
run;

**** TRANSPOSE THE NORMALIZED SBP VALUES TO A FLAT STRUCTURE.;
proc transpose
   data = sbp
   out = sbpflat
   prefix = VISIT;
      by subject;
      id visit;
      var sbp;
run;
```

The resulting data set, "sbpflat," looks like this:

Obs	subject _id	_NAME_	VISIT1	VISIT2	VISIT3	VISIT4	VISIT5
1	101	sbp	160	150	140	130	120
2	202	sbp	141	151	161	171	181

You can see that the systolic blood pressures have been transposed from row-based data to column-based data and that the "_NAME_" variable is superfluous and may be dropped. A common trap with PROC TRANSPOSE occurs when there are missing data. For example, look at the following program, where the row for visit 2 for subject 101 is dropped.

Program 4.6 Using PROC TRANSPOSE Without and With an ID Statement

```
**** INPUT SAMPLE NORMALIZED SYSTOLIC BLOOD PRESSURE VALUES.
**** SUBJECT = PATIENT NUMBER, VISIT = VISIT NUMBER,
**** SBP = SYSTOLIC BLOOD PRESSURE.;
   data sbp;
   input subject $ visit sbp;
   datalines;
   101 1 160          Notice missing visit = 2
   101 3 140
   101 4 130
   101 5 120
   202 1 141
   202 2 151
   202 3 161
   202 4 171
   202 5 181
   ;
run;

**** TRANSPOSE THE NORMALIZED SBP VALUES TO A FLAT STRUCTURE.;
proc transpose
   data = sbp
   out = sbpflat
   prefix = VISIT;
      by subject;
      var sbp;
run;
```

The resulting data set, "sbpflat," looks like this:

Obs	subject	_NAME_	VISIT1	VISIT2	VISIT3	VISIT4	VISIT5
1	101	sbp	160	140	130	120	.
2	202	sbp	141	151	161	171	181

In this example you can see that the transposition for subject 101 is incorrect because of the missing record for visit 2. In this case, the systolic blood pressure values are shifted down one slot so that visit 3 data are mistakenly placed in the visit 2 slot, and so forth for visit 4 and visit 5. To remedy this error, use an ID statement in PROC TRANSPOSE as follows:

```
proc transpose
   data = sbp
   out = sbpflat
   prefix = VISIT;
      by subject;
      id visit;
      var sbp;
run;
```

The resulting data set, "sbpflat," looks like this:

Obs	Subject	_NAME_	VISIT1	VISIT2	VISIT3	VISIT4	VISIT5
1	101	sbp	160	.	140	130	120
2	202	sbp	141	151	161	171	181

You can see that the missing systolic blood pressure value for visit 2 is now properly assigned for subject 101. If order is important when transposing row data to columns, then the use of an ID statement in PROC TRANSPOSE is imperative. Otherwise, your data will silently shift to the left in the transposed file.

There may be times when a DATA step with arrays is a better means to transpose data. This is true when the data to be transposed have more than one record per BY group variable or when there is a need to have the resulting data set include data that are not in the source data set. In clinical trials missing data is a very common issue. Let's look at a derivation of the previous systolic blood pressure transposition problem where visit 2 is always missing.

```
**** INPUT SAMPLE NORMALIZED SYSTOLIC BLOOD PRESSURE VALUES.
**** SUBJECT = PATIENT NUMBER, VISIT = VISIT NUMBER,
**** SBP = SYSTOLIC BLOOD PRESSURE.;
data sbp;
input subject $ visit sbp;
datalines;
101 1 160
101 3 140          ◄──────  Notice missing visit = 2
101 4 130
101 5 120
202 1 141
202 3 161          ◄──────  Notice missing visit = 2
202 4 171
202 5 181
;
run;

**** TRANSPOSE THE NORMALIZED SBP VALUES TO A FLAT STRUCTURE.;
proc transpose
    data = sbp
    out = sbpflat
    prefix = VISIT;
        by subject;
        id visit;
        var sbp;
run;
```

The resulting data set, "sbpflat," looks like this:

Obs	subject	_NAME_	VISIT1	VISIT3	VISIT4	VISIT5
1	101	sbp	160	140	130	120
2	202	sbp	141	161	171	181

Notice the missing column for visit 2. This is exactly what you would expect PROC TRANSPOSE to give you. PROC TRANSPOSE transposed the data that were present and could not be expected to know about visits that are not represented in the data. However, often in clinical trials reporting you want to report on all visits, treatments, or other expected parameters whether they are represented in the actual data or not. In this case, a DATA step with arrays is a better choice to transform the data. Here is an example of the previous transposition that includes all visits 1–5, regardless of which visits are included in the underlying data.

Program 4.7 Transposing Data with the DATA Step

```
**** INPUT SAMPLE NORMALIZED SYSTOLIC BLOOD PRESSURE VALUES.
**** SUBJECT = PATIENT NUMBER, VISIT = VISIT NUMBER,
**** SBP = SYSTOLIC BLOOD PRESSURE.;
data sbp;
input subject $ visit sbp;
datalines;
101 1 160
101 3 140            ◄───  Notice missing visit = 2
101 4 130
101 5 120
202 1 141            ◄───  Notice missing visit = 2
202 3 161
202 4 171
202 5 181
;
run;

**** SORT SBP VALUES BY SUBJECT.;
proc sort
   data = sbp;
      by subject;
run;

**** TRANSPOSE THE NORMALIZED SBP VALUES TO A FLAT STRUCTURE.;
data sbpflat;
   set sbp;
      by subject;

      keep subject visit1-visit5;
      retain visit1-visit5;

      **** DEFINE ARRAY TO HOLD SBP VALUES FOR 5 VISITS.;
      array sbps {5} visit1-visit5;

      **** AT FIRST SUBJECT, INITIALIZE ARRAY TO MISSING.;
      if first.subject then
         do i = 1 to 5;
            sbps{i} = .;
         end;

      *** AT EACH VISIT LOAD THE SBP VALUE INTO THE PROPER SLOT
      **** IN THE ARRAY.;
      sbps{visit} = sbp;
```

```
          **** KEEP THE LAST OBSERVATION PER SUBJECT WITH 5 SBPS.;
          if last.subject;
run;
```

The resulting data set, "sbpflat," looks like this:

Obs	subject_id	visit1	visit2	visit3	visit4	visit5
1	101	160	.	140	130	120
2	202	141	.	161	171	181

You can see that although visit 2 never actually occurred in the data, it is represented as a column in the final data set.

PROC TRANSPOSE is an efficient way to transpose a SAS data set when the transposition process is simple. However, if the transposition process is more complicated, and involves transforming to a data set where all possible columns must be represented or where there are multiple records per BY group, then a DATA step with arrays is probably a better choice.

Categorical Data and Why Zero and Missing Results Differ Greatly

A key concept to understand when deriving categorical analysis data set variables is that a missing value result is entirely different from a zero result. This is one of the most common problems when dealing with categorical analysis variable specifications, because the "missing versus zero" problem is overlooked. Here is a simple table to show the difference in definition between a categorical variable missing and zero result:

When a SAS Categorical Value Is...	
Missing (.), it means...	Zero (0), it means...
• The response is unknown.	• The response is known.
• The observation will not be included in population analysis and denominator definitions.	• The response is "No" when the categorical variable is a Boolean variable.
	• The observation will be included in population analysis and denominator definitions.

There are negative consequences when a zero result is assumed for a categorical variable. When a zero result is assumed, inferential analysis can provide an incorrect result and descriptive statistics can be skewed.

Let's look at an example of what might happen in the following data set of patient deaths when we assume a zero result in place of a missing result. Here is a data set of subject death information (death) by treatment (trt), where death = 0 if a patient did not die and death = 1 if the patient did die. If we are not certain that a patient died, then we assume that the patient did not die and use death = 0.

```
****  INPUT SAMPLE MORTALITY DATA.
****  SUBJECT = PATIENT NUMBER, DEATH = 1=DIED,0=IF ALIVE,
****  TRT = STUDY TREATMENT.;
data death;
input subject death trt $ @@;
datalines;
101 0 a 102 0 b 103 1 a 104 0 b 105 1 a 106 0 b 107 1 a 108 0 b
109 1 a 110 1 b 111 1 a 112 1 b 113 0 a 114 0 b 115 1 a 116 0 b
117 1 a 118 0 b 119 1 a 120 1 b 121 1 a 122 1 b 123 1 a 124 1 b
125 1 a 126 0 b 127 1 a 128 0 b 129 1 a 130 0 b 131 1 a 132 1 b
133 1 a 134 1 b 135 1 a 136 1 b 137 1 a 138 1 b 139 1 a 140 1 b
201 0 b 202 1 a 203 0 b 204 0 a 205 1 b 206 0 a 207 1 b 208 1 a
209 1 b 210 1 a 211 1 b 212 1 a 213 0 b 214 1 a 215 0 b 216 0 a
217 1 b 218 0 a 219 1 b 220 1 a 221 1 b 222 1 a 223 1 b 224 1 a
225 0 b 226 1 a 227 0 b 228 0 a 229 1 b 230 0 a 231 1 b 232 1 a
233 1 b 234 1 a 235 1 b 236 1 a 237 0 b 238 1 a 239 0 b 240 0 a
;
run;


proc freq
   data = death;
      tables death * trt /norow nocol chisq;
run;
```

This code produces the following partial output from PROC FREQ:

Table of death by trt			
death	**trt**		
Frequency Percent	**a**	**b**	**Total**
0	9 11.25	18 22.50	27 33.75
1	31 38.75	22 27.50	53 66.25
Total	40 50.00	40 50.00	80 100.00

Statistics for Table of death by trt

Statistic	DF	Value	Prob
Chi-Square	1	4.5283	0.0333

You see that all 80 patients have their mortality status reported and that just by scanning the frequency tabulation you can tell that more subjects receiving treatment "a" have died. You also see from the chi-square *p*-value of 0.0333 that treatment "a" has significantly more deaths at an alpha level of .05.

Now, what if you discovered that in some cases you really did not know if some patients had died? In fact, you were assuming death = 0 as a default for all patients. The corrected database looks like this, with the changes underlined to indicate what is unknown:

```
**** INPUT SAMPLE MORTALITY DATA.
**** SUBJECT = PATIENT NUMBER, DEATH = 1=DIED,0=IF ALIVE,
**** TRT = STUDY TREATMENT.;
data death;
input subject death trt $ @@;
datalines;
101 0 a 102 . b 103 1 a 104 . b 105 1 a 106 . b 107 1 a 108 . b
109 1 a 110 1 b 111 1 a 112 1 b 113 0 a 114 0 b 115 1 a 116 0 b
117 1 a 118 0 b 119 1 a 120 1 b 121 1 a 122 1 b 123 1 a 124 1 b
125 1 a 126 0 b 127 1 a 128 0 b 129 1 a 130 0 b 131 1 a 132 1 b
133 1 a 134 1 b 135 1 a 136 1 b 137 1 a 138 1 b 139 1 a 140 1 b
```

```
201 0 b 202 1 a 203 0 b 204 0 a 205 1 b 206 0 a 207 1 b 208 1 a
209 1 b 210 1 a 211 1 b 212 1 a 213 0 b 214 1 a 215 0 b 216 0 a
217 1 b 218 0 a 219 1 b 220 1 a 221 1 b 222 1 a 223 1 b 224 1 a
225 0 b 226 1 a 227 0 b 228 0 a 229 1 b 230 0 a 231 1 b 232 1 a
233 1 b 234 1 a 235 1 b 236 1 a 237 0 b 238 1 a 239 0 b 240 0 a
;
run;

proc freq
   data = death;
      tables death*trt /norow nocol chisq;
run;
```

This code produces the following partial output from PROC FREQ:

Table of death by trt			
death	**trt**		
Frequency Percent	**a**	**b**	**Total**
0	9 11.84	14 18.42	23 30.26
1	31 40.79	22 28.95	53 69.74
Total	40 52.63	36 47.37	76 100.00
Frequency Missing = 4			

Statistics for Table of death by trt

Statistic	DF	Value	Prob
Chi-Square	1	2.4114	0.1205

You see that by changing only four deaths from the "no" result to missing, you went from a significant finding (that treatment "a" is associated with more deaths) to a finding in which the association is not significant. Also note that percentage calculations may be affected as well, because when a value is zero it gets counted toward a denominator, but when it is missing is often gets omitted from the denominator for percentage calculations.

The "missing versus zero" problem is a chronic issue in clinical trial analysis and reporting. You will save considerable time by specifying categorical analysis variables with the "missing versus zero" problem in mind before you begin analysis data set programming. You should ask whether a missing value is a valid result for any analysis variable.

Performing Many-to-Many Comparisons/Joins

At times it is necessary to merge two data sets not simply with a one-to-one or one-to-many merge, but with a many-to-many merge. This task generally cannot be accomplished properly with a simple MERGE-BY statement combination. You can perform complicated merges with advanced SET statement processing, but in this case the best tool to employ for many-to-many merges is an SQL join. Let's look at an SQL-based example where we want to merge two data sets based on a range of values in a many-to-many relationship.

Imagine you have a data set of adverse event data and a data set of concomitant medications, and you want to know if a concomitant medication was given to a patient during the time of the adverse event. The following program defines the two data sets and joins them with PROC SQL so that you get all medications taken during any specific adverse event.

Program 4.8 Performing a Many-to-Many Join with PROC SQL

```
**** INPUT SAMPLE ADVERSE EVENT DATA.
**** SUBJECT = PATIENT NUMBER, AE_START = START DATE OF AE,
**** AE_STOP = STOP DATE OF AE, ADVERSE_EVENT = NAME OF EVENT.;
data aes;
informat ae_start date9. ae_stop date9.;
input @1  subject $3.
      @5  ae_start date9.
      @15 ae_stop date9.
      @25 adverse_event $15.;
datalines;
101 01JAN2004 02JAN2004 Headache
101 15JAN2004 03FEB2004 Back Pain
102 03NOV2003 10DEC2003 Rash
102 03JAN2004 10JAN2004 Abdominal Pain
102 04APR2004 04APR2004 Constipation
;
run;

**** INPUT SAMPLE CONCOMITANT MEDICATION DATA.
**** SUBJECT = PATIENT NUMBER, AE_START = START DATE OF AE,
**** AE_STOP = STOP DATE OF AE, ADVERSE_EVENT = NAME OF EVENT.;
data conmeds;
```

```
informat cm_start date9. cm_stop date9.;
input @1  subject $3.
      @5  cm_start date9.
      @15 cm_stop date9.
      @25 conmed $20.;
datalines;
101 01JAN2004 01JAN2004 Acetaminophen
101 20DEC2003 20MAR2004 Tylenol w/ Codeine
101 12DEC2003 12DEC2003 Sudafed
102 07DEC2003 18DEC2003 Hydrocortisone Cream
102 06JAN2004 08JAN2004 Simethicone
102 09JAN2004 10MAR2004 Esomeprazole
;
run;

**** MERGE/JOIN ADVERSE EVENTS WITH CONCOMITANT MEDICATIONS.
**** KEEP MEDICATIONS THAT STARTED OR STOPPED DURING AN ADVERSE
**** EVENT OR ENTIRELY SPANNED ACROSS AN ADVERSE EVENT.;
proc sql;
   create table ae_meds as
   select a.subject, a.ae_start, a.ae_stop,
          a.adverse_event, c.cm_start, c.cm_stop,
          c.conmed from
   aes as a  left join  conmeds as c
   on (a.subject = c.subject) and
      ( (a.ae_start <= c.cm_start <= a.ae_stop) or
        (a.ae_start <= c.cm_stop <= a.ae_stop) or
       ((c.cm_start < a.ae_start) and (a.ae_stop < c.cm_stop)));
quit;
```

This code produces the following "ae_meds" data set:

Obs	subject	ae_start	ae_stop	adverse_event	cm_start	cm_stop	conmed
1	101	01JAN2004	02JAN2004	Headache	01JAN2004	01JAN2004	Acetaminophen
2	101	01JAN2004	02JAN2004	Headache	20DEC2003	20MAR2004	Tylenol w/ Codeine
3	101	15JAN2004	03FEB2004	Back Pain	20DEC2003	20MAR2004	Tylenol w/ Codeine
4	102	03JAN2004	10JAN2004	Abdominal Pain	09JAN2004	10MAR2004	Esomeprazole
5	102	03JAN2004	10JAN2004	Abdominal Pain	06JAN2004	08JAN2004	Simethicone
6	102	03NOV2003	10DEC2003	Rash	07DEC2003	18DEC2003	Hydrocortisone Cream
7	102	04APR2004	04APR2004	Constipation	.	.	

Here you can see that each adverse event that was experienced has now been merged with each concomitant medication taken during that time. This is due to the SQL SELECT statement ON clause, which allows for complicated logic to be applied to the join.

Using SQL, you can create very sophisticated database joins that were not easily done before SQL became part of Base SAS. Keep in mind that these data set joins can get very complex. For example, the previous scenario gets quite a bit more complicated with the introduction of missing start and stop dates, which are a common occurrence in clinical trial data.

Using Medical Dictionaries

Medical dictionaries often need to be referenced when creating various analysis data sets. For instance, perhaps the raw adverse event database in your clinical data management system contains only the MedDRA code. The code is worth having, but you would need the adverse event body system and preferred medical term to provide a useful summary of events.

There are hundreds of proprietary medical dictionaries in circulation. Fortunately, a number of medical dictionaries are headed for retirement. The pharmaceutical industry along with regulators and standards groups is actively working to reduce the number of coding dictionaries. So we will discuss MedDRA for adverse events and WHODrug for medications, because they are the two most commonly used medical dictionaries in clinical trials today.

Medical Dictionary for Regulatory Activities (MedDRA)

The Medical Dictionary for Regulatory Activities (MedDRA) is a creation of the International Conference on Harmonization, and it is used to categorize and code diseases, disorders, and adverse events. The five levels and associated codes to the MedDRA coding hierarchy are as follows:

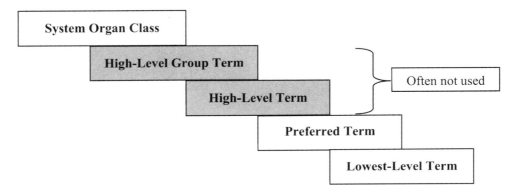

This MedDRA illustration begins at the bottom with the lowest-level term. This is the term used to describe medical conditions that your clinical data management group typically assigns to a given event. The lowest-level term then rolls up into a preferred term; many low-level terms may map to a single preferred term. In general, the high level term and high-level group terms are not used for clinical trial reporting. If needed, these high-level terms provide a way to roll up events to a more general level of description without going all the way to body system. The top-most system organ class, or body system, level represents the body system affected by the event.

The essential pieces of MedDRA to have in SAS are the preferred term and body system, as this is what most clinical summaries and analyses require. The following is a piece of simplified SAS code that brings the lowest-level terms, preferred terms, and body systems together into one data set.

Program 4.9 Bringing the MedDRA Dictionary Tables Together

```
**** SORT LOW LEVEL TERM DATA FROM MEDDRA WHERE
**** LOW_LEVEL_TERM = LOWEST LEVEL TERM, LLT_CODE = LOWEST
**** LEVEL TERM CODE, AND PT_CODE = PREFERRED TERM CODE.;
proc sort
    data = low_level_term(keep = low_level_term llt_code
            pt_code);
        by pt_code;
run;


**** SORT PREFERRED TERM DATA FROM MEDDRA WHERE
**** PREFERRED_TERM = PREFERRED TERM, SOC_CODE = SYSTEM
**** ORGAN CLASS CODE, AND PT_CODE = PREFERRED TERM CODE.;
proc sort
    data = preferred_term(keep = preferred_term pt_code soc_code);
        by pt_code;
run;


**** MERGE LOW LEVEL TERMS WITH PREFERRED TERMS KEEPING ALL LOWER
**** LEVEL TERM RECORDS.;
data llt_pt;
    merge low_level_term (in = inlow)
            preferred_term;
        by pt_code;

        if inlow;
run;


**** SORT BODY SYSTEM DATA FROM MEDDRA WHERE
**** SYSTEM_CLASS_TERM = SYSTEM ORGAN CLASS TERM AND SOC_CODE =
**** SYSTEM ORGAN CLASS CODE.;
proc sort
    data = soc_term(keep = system_class_term soc_code);
        by soc_code;
run;


**** SORT LOWER LEVEL TERM AND PREFERRED TERMS FOR MERGE WITH
**** SYSTEM ORGAN CLASS DATA.;
proc sort
    data = llt_pt;
        by soc_code;
run;
```

```
**** MERGE PREFERRED TERM LEVEL WITH BODY SYSTEMS;
data meddra;
   merge llt_pt (in = in_llt_pt)
         soc_term;
      by soc_code;

      if in_llt_pt;
run;
```

The final "meddra" data set in this program contains the lower-level term code (llt_code) that can then be merged with the adverse events or medical conditions database. By merging the MedDRA dictionary data with the disease data, you can match the verbatim event text captured on the case report form with the preferred term and associated body system. Then you can summarize these data by body system and preferred term; you will see an example of this in Chapter 5.

There are a couple of items to note about MedDRA and the preceding sample SAS code. First, the example assumes that you have already managed to read the system organ class, preferred term, and lowest-level term data tables from the MedDRA dictionary into SAS. Second, the preferred terms may map to more than one system organ class. In this case, MedDRA provides a variable at the preferred term level to indicate what the primary system organ class is. So, you may need to subset the preferred term data set in the example in order to select just the primary system organ class.

World Health Organization Drug Dictionary (WHODrug)

The World Health Organization Drug Dictionary (WHODrug) is a creation of the Uppsala Monitoring Centre in Sweden. WHODrug is designed to take international, proprietary, and non-proprietary or generic drug names and classify them as common preferred terms. It is with these preferred terms that useful statistical summaries can be provided.

These are the critical fields from the WHODrug medicinal product data file:

Field	Description
Drug record number	6-byte field identifying a medicinal product.
Sequence number 1	2-byte field distinguishing salts or esters. A value of "01" indicates the base substance without any salt or ester.
Sequence number 2	3-byte field distinguishing trade names for medicinal product. A value of "001" indicates the preferred term for the drug.
Drug name	80-byte field identifying a medicinal product name.

In order to obtain the preferred term for a given drug, you have to pull drug record numbers, where sequence number 1 is "01" and sequence number 2 is "001." Also, redundancies in the resulting data set must be stripped out with first-dot processing or a NODUPKEY on a PROC SORT. The following SAS code shows how to get preferred terms from WHODrug.

Program 4.10 Pulling Preferred Terms out of WHODrug

```
proc sort
    data = whodrug(keep = seq1 seq2 drug_name drugrecno
                   where = (seq1 = '01' and seq2 = '001') )
          nodupkey;
       by drugrecno drug_name;
run;
```

The data set in this program can then be merged with the medications data from the case report form by the "drugrecno" variable to attach the preferred term from the WHODrug dictionary with the verbatim drug name collected from the case report form.

WHODrug also provides a drug classification higher than the preferred term, called the Anatomical-Therapeutic-Chemical (ATC) classification. The ATC classification allows for categorizing of drug data by anatomical, therapeutic, or chemical groupings. The ATC classification would allow you to summarize medications by a high-level common drug class. Unfortunately, a given drug can appear in multiple ATC classes. Because of this, the ATC classifications should be used with care, because you generally do not want to count a drug more than once for a patient. However, the ATC classification can be useful if you want to summarize or provide a list of protocol-prohibited medications.

Other Tricks and Traps in Data Manipulation

What We Learned from Y2K

It seems like just yesterday when the world was going to end at the start of the new century because we wanted to save disk space by using two bytes to store years. The world did not end, but you would be remiss if you were to forget the problem of Y2K and date storage. All software, including SAS, must impute a century when you do not specify one. Therefore, if you want to make sure the century is assigned correctly, then always explicitly state the century when working with dates.

When a century is imputed by software, it is usually done based on something called a pivot point. The pivot point in SAS for determining the 100-year window is based on the YEARCUTOFF option, which is set by default to 1920. Let's look at an example to see what this means. The following SAS code uses both implicit and explicit century dates right at the YEARCUTOFF pivot point of 1920.

Program 4.11 Using Implicit or Explicit Centuries with Dates

```
**** DISPLAY YEARCUTOFF SETTING PIVOT POINT;
proc options option = yearcutoff;
run;

**** DATES DEFINED WITH IMPLICIT CENTURY;
data _null_;
   date = "01JAN19"d;
   put date = date9.;
   date = "01JAN20"d;
   put date = date9.;
run;

**** DATES DEFINED WITH EXPLICIT CENTURY;
data _null_;
   date = "01JAN1919"d;
   put date = date9.;
   date = "01JAN1920"d;
   put date = date9.;
run;
```

This program produces the following SAS log:

```
1    proc options option = yearcutoff;
2    run;

     SAS (r) Proprietary Software Release 9.1   TS1M0

YEARCUTOFF=1920    Cutoff year for DATE and DATETIME informats
and functions

3
4    **** IMPLICIT CENTURY;
5    data _null_;
6      date = "01JAN19"d;
7      put date=date9.;
8      date = "01JAN20"d;
9      put date=date9.;
10   run;

date=01JAN2019
date=01JAN1920

11
12   **** EXPLICIT CENTURY;
13   data _null_;
```

```
14      date = "01JAN1919"d;
15      put date=date9.;
16      date = "01JAN1920"d;
17      put date=date9.;
18   run;

date=01JAN1919
date=01JAN1920
```

Note that the implicit century DATA step produces dates in two different centuries, but
that when you explicitly state the century there is no subsequent century confusion. You
can define YEARCUTOFF as needed for your specific SAS applications, but it is wise to
have a reasonable system-wide YEARCUTOFF default set. Also, if you use explicit
centuries whenever possible, you minimize the implicit century risk.

As an aside, Microsoft Excel's pivot-point year is 1930 by default. So, if you enter an
implicit century date in Excel such as "01/01/29," it understands that as 01/01/2029, but
"01/01/30" is understood as 01/01/1930. This is useful to know when implicit century
data pass through Excel into SAS in some fashion.

Redefining an Already Existing Data Set Variable

Often you want to redefine an already existing variable within a SAS DATA step. As
simple as this may sound, it can lead to unexpected results if not done carefully. The
following example displays some unexpected behavior that may occur when you redefine
a variable within a DATA step. In this example you want to flag the subject who had the
"Fatal MI" adverse event as having died (death = 1).

Program 4.12 Redefining a Variable within a DATA Step

```
****  INPUT SAMPLE ADVERSE EVENT DATA WHERE SUBJECT = PATIENT ID
****  AND ADVERSE_EVENT = ADVERSE EVENT TEXT.;
data aes;
input @1   subject $3.
      @5   adverse_event $15.;
datalines;
101 Headache
102 Rash
102 Fatal MI
102 Abdominal Pain
102 Constipation
;
run;
```

```
**** INPUT SAMPLE DEATH DATA WHERE SUBJECT = PATIENT NUMBER AND
**** DEATH = 1 IF PATIENT DIED, 0 IF NOT.;
data death;
input @1 subject $3.
      @5 death 1.;
datalines;
101 0
102 0
;
run;

**** SET DEATH = 1 FOR PATIENTS WHO HAD ADVERSE EVENTS THAT
**** RESULTED IN DEATH.;
data aes;
   merge demog aes;
      by subject;

      if adverse_event = "Fatal MI" then
         death = 1;
run;

proc print
   data = aes;
run;
```

This program produces the following unexpected output:

Obs	subject	adverse_event	death
1	101	Headache	0
2	102	Rash	0
3	102	Fatal MI	1
4	102	Abdominal Pain	1
5	102	Constipation	1

Notice how the "Abdominal Pain" and "Constipation" observations also were flagged as death = 1. This happens because, by design, SAS automatically retains the values of all variables brought into a DATA step via the MERGE, SET, or UPDATE statement. In this example, because the "death" variable already exists in the "death" data set, that variable gets retained. When you assign death = 1 for "Fatal MI," the subsequent records within that BY group are also set to 1. You should carefully read "Combining SAS Data Sets" in the Base SAS documentation for a thorough explanation of why this happens.

For the previous example, the following code is a workaround to the automatic variable retention feature.

```
**** FLAG EVENTS THAT RESULTED IN DEATH;
data aes;
   merge death(rename = (death = _death)) aes;
      by subject;

      **** DROP OLD DEATH VARIABLE.;
      drop _death;

      **** CREATE NEW DEATH VARIABLE.;
      if adverse_event = "Fatal MI" then
        death = 1;
      else
        death = _death;
run;

proc print;
run;
```

This program produces the following output:

Obs	subject_id	adverse_event	death
1	101	Headache	0
2	102	Rash	0
3	102	Fatal MI	1
4	102	Abdominal Pain	0
5	102	Constipation	0

Now you see that death = 1 only for the "Fatal MI" as desired. This was accomplished by changing the name of the "death" variable to "_death" on the way into the "aes" DATA step and then using the "_death" variable in defining a newly created "death" variable. Finally, "_death" is dropped from the outgoing copy of the "aes" data set.

Even after reading the SAS documentation "Combining SAS Data Sets" and understanding how the program data vector works, it can still be a bit confusing as to when it may be safe to redefine a pre-existing data set variable in place. The good news is that there is a safe and simple way to avoid the unplanned retention of variables: Do not redefine a pre-existing variable within a DATA step, and use a workaround like the one shown in the previous code.

Performing Floating-Point Comparisons

At times you need to perform some form of floating-point number comparison. The following example illustrates such a comparison where you are setting laboratory value flags to indicate whether a lab test is above or below normal.

Program 4.13 Using the ROUND Function with Floating-Point Comparisons

```
**** FLAG LAB VALUE AS LOW OR HIGH;
data labs;
   set labs;

      if .z < lab_value < 3.15 then
         hi_low_flag = "L";
      else if lab_value > 5.5 then
         hi_low_flag = "H";
   run;
```

At first glance the SAS code snippet above looks fine. If the lab value is less than 3.15 then the lab value is low, and if it is greater than 5.5 then the lab value is high. The problem is that floating-point numbers are stored on computers as approximations of the value that you may see in a SAS printout. In other words, 3.15 might be stored as 3.15000000000000000000000000000000012, which might make the program snippet above not work as hoped. To handle the floating-point comparison problem, this example can be written more safely like this:

```
**** FLAG LAB VALUE AS LOW OR HIGH;
data labs;
   set labs;

      if .z < round(lab_value,.000000001) < 3.15 then
         hi_low_flag = "L";
      else if round(lab_value,.000000001) > 5.5 then
         hi_low_flag = "H";
   run;
```

Notice how the "lab_value" variable has been rounded to an arbitrary precision to the ninth decimal place. This gets around the floating-point comparison problem, while at the same time it avoids rounding the "lab_value" variable to a precision where meaningful data are lost.

You should read Technical Support Note "TS-230: Dealing with Numeric Representation Error in SAS Applications" to learn more about SAS floating-point numbers and storage precision in SAS. Another good resource for rounding issues is Ron Cody's *SAS Functions by Example* (SAS Press, 2004). In short, whenever you perform comparisons on numbers that are not integers, you should consider using the ROUND function.

Common Analysis Data Sets

In this section we take the aforementioned principles and guidelines for analysis data sets and apply them to creating the most common analysis data sets. The critical variables, change-from-baseline, and time-to-event data sets are presented. Although these are the most common analysis data sets that a statistical programmer will encounter, they are by no means all of the possible analysis data sets. When it comes to analysis data sets, there is no limit to the diversity of data that you may have to create.

Critical Variables Data Set

The first analysis data set that should be constructed for any clinical trial analysis is the data set of critical variables. The critical variables typically include demographic variables (age, sex, race), population status variables (intent-to-treat, per-protocol, other analysis population flags), and a study treatment assignment variable. Any other variable that might be used to stratify a population for analyses should probably go in the critical variables data set as well. The critical variables analysis data set always has a single observation per subject to simplify the process of merging with other data sets. The whole purpose of the critical variables data set is to capture in one place the essential analysis stratification variables that are used throughout the statistical analysis and reporting.

Unfortunately, it is often hard to know a critical variables data set when you see it. It may be named "pop," "critvar," "essential," "patients," "demog," or even "dm," so it can be a bit hard to find. When creating this critical variables data set, you should try to create a good descriptive name for it.

Change-from-Baseline Data Set

The purpose of using change-from-baseline analysis data sets is to measure what effect some therapeutic intervention had on some kind of diagnostic measure. A measure is taken before and after therapy, and a difference and sometimes a percentage difference are calculated for each post-baseline measure. These data sets are generally normalized vertical structure data sets. Here is an example of how such a data set could be created for systolic and diastolic blood pressure data.

Program 4.14 Creating a Blood Pressure Change-from-Baseline Data Set

```
**** INPUT SAMPLE BLOOD PRESSURE VALUES WHERE
**** SUBJECT = PATIENT NUMBER, WEEK = WEEK OF STUDY, AND
**** TEST = SYSTOLIC (SBP) OR DIASTOLIC (DBP) BLOOD PRESSURE.;
data bp;
input subject $ week test $ value;
datalines;
101 0 DBP 160
101 0 SBP  90
101 1 DBP 140
101 1 SBP  87
101 2 DBP 130
101 2 SBP  85
101 3 DBP 120
101 3 SBP  80
202 0 DBP 141
202 0 SBP  75
202 1 DBP 161
202 1 SBP  80
202 2 DBP 171
202 2 SBP  85
202 3 DBP 181
202 3 SBP  90
;
run;

**** SORT DATA BY SUBJECT, TEST NAME, AND WEEK;
proc sort
   data = bp;
      by subject test week;
run;

**** CALCULATE CHANGE FROM BASELINE SBP AND DBP VALUES.;
data bp;
   set bp;
      by subject test week;

      **** CARRY FORWARD BASELINE RESULTS.;
      retain baseline;
      if first.test then
         baseline = .;
```

```
         **** DETERMINE BASELINE OR CALCULATE CHANGES.;
         if visit = 0 then
            baseline = value;
         else if visit > 0 then
            do;
               change = value - baseline;
               pct_chg = ((value - baseline) /baseline )*100;
            end;
run;

proc print
   data = bp;
run;
```

This program produces the following output:

Obs	subject	week	test	value	baseline	change	pct_chg
1	101	0	DBP	160	160	.	.
2	101	1	DBP	140	160	-20	-12.5000
3	101	2	DBP	130	160	-30	-18.7500
4	101	3	DBP	120	160	-40	-25.0000
5	101	0	SBP	90	90	.	.
6	101	1	SBP	87	90	-3	-3.3333
7	101	2	SBP	85	90	-5	-5.5556
8	101	3	SBP	80	90	-10	-11.1111
9	202	0	DBP	141	141	.	.
10	202	1	DBP	161	141	20	14.1844
11	202	2	DBP	171	141	30	21.2766
12	202	3	DBP	181	141	40	28.3688
13	202	0	SBP	75	75	.	.
14	202	1	SBP	80	75	5	6.6667
15	202	2	SBP	85	75	10	13.3333
16	202	3	SBP	90	75	15	20.0000

Note that the new "change" and "pct_change" variables have been added to the data set for absolute change from baseline and percentage change from baseline, respectively. Percentage change from baseline is considered a useful measure because it gives a change value adjusted for baseline value. Also note that the baseline observations (week = 0) have been retained in this data set because they are often useful to keep when it comes time for data summarization.

Time-to-Event Data Set

A time-to-event analysis data set captures the information about the time distance between therapeutic intervention and some other particular event. There are two time-to-event analysis variables that deserve special attention and definition. They are as follows:

Variable	Definition
Event/Censor	A binomial outcome such as "success/failure," "death/life," "heart attack/no heart attack." If the event happened to the subject, then the event variable is set to 1. If it is certain that the patient did not experience the event, then the event variable is set to 0. Otherwise, this variable should be missing.
Time to Event	This variable captures the time (usually study day) from therapeutic intervention to the event date or censor date. If the event occurred for a subject, the time to event is the study day at that event. If the event did not occur, then the time to event is set to the censor date that is often the last known follow-up date for a subject.

So for every clinical event of concern there is an event binomial flag and a time-to-event variable. Time-to-event data sets are typically represented in a flat denormalized single observation per subject data set.

Note that the term "censor" is introduced in the preceding table. The log-rank test (invoked in SAS with PROC LIFETEST) and the Cox proportional hazards model (invoked in SAS with PROC PHREG) allow for censoring observations in a time-to-event analysis. These tests adjust for the fact that at some point a patient may no longer be able to experience an event. The censor date is the last known time that the patient did not experience a given event and the point at which the patient is no longer considered able to experience the event. Often the censor date is the last known date of patient follow-up, but a patient could be censored for other reasons, such as having taken a protocol-prohibited medication.

Creating time-to-event data sets can be a difficult programming task, especially during interim data analyses, such as for a DSMB. This is usually because the event data itself are captured in more than one place in the case report form and the censor date may be difficult to obtain. For example, perhaps the event of interest is death. You may have to search the adverse events CRF page, the study termination CRF page, clinical endpoint committee CRFs, and perhaps a special death events CRF page just to gather all of the known death events and dates. For subjects who did not experience the event of interest, you may not have a study termination form to provide the censoring date, so you may have to use some surrogate data to create a censor date.

The following is a very simplistic example of what a time-to-event data set might look like when the event of interest is seizure. Here you assume that there is a seizure event form that collects whether a subject had a seizure and the date when the seizure occurred. You also assume that you do not need to search other ancillary data forms such as adverse events for seizure events.

Program 4.15 Creating a Time-to-Event Data Set for Seizures

```
**** INPUT SAMPLE SEIZURE DATA WHERE
**** SUBJECT = PATIENT NUMBER, SEIZURE = BOOLEAN FLAG
**** INDICATING A SEIZURE AND SEIZDATE = DATE OF SEIZURE.;
data seizure;
informat seizdate date9.;
format seizdate date9.;
label subject  = "Patient Number"
      seizdate = "Date of Seizure"
      seizure  = "Seizure: 1=Yes,0=No";
input subject $ seizure seizdate;
datalines;
101 1 05MAY2004
102 0 .
103 . .
104 1 07JUN2004
;
run;

**** INPUT SAMPLE END OF STUDY DATA WHERE
**** SUBJECT = PATIENT NUMBER, EOSDATE = END OF STUDY DATE.;
data eos;
informat eosdate date9.;
format eosdate date9.;
label subject  = "Patient Number"
      eosdate = "End of Study Date";
input subject $ eosdate;
datalines;
101 05AUG2004
```

```
102 10AUG2004
103 12AUG2004
104 20AUG2004
;
run;

**** INPUT SAMPLE DOSING DATA WHERE
**** SUBJECT = PATIENT NUMBER AND DOSEDATE = DRUG DOSING DATE.;
data dosing;
informat dosedate date9.;
format dosedate date9.;
label subject  = "Patient Number"
      dosedate = "Start of Drug Therapy";
input subject $ dosedate;
datalines;
101 01JAN2004
102 03JAN2004
103 06JAN2004
104 09JAN2004
;
run;

**** CREATE TIME TO SEIZURE DATA SET;
data time_to_seizure;
   merge dosing eos seizure;
      by subject;

      if seizure = 1 then
         time_to_seizure = seizdate - dosedate + 1;
      else if seizure = 0 then
         time_to_seizure = eosdate - dosedate + 1;
      else
         time_to_seizure = .;

      label time_to_seizure = "Days to Seizure or Censor Day";
run;

proc print
  label data = time_to_seizure;
run;
```

This program produces the following output:

Newly Derived

Obs	Start of Drug Therapy	Patient Number	End of Study Date	Date of Seizure	Seizure: 1=Yes,0=No	Days to Seizure or Censor Day
1	01JAN2004	101	05AUG2004	05MAY2004	1	126
2	03JAN2004	102	10AUG2004	.	0	221
3	06JAN2004	103	12AUG2004	.	.	.
4	09JAN2004	104	20AUG2004	07JUN2004	1	151

Note that if there were many seizures for a subject, you would have wanted only the time to the first seizure presented, as this is a single record per subject data set and analysis.

Chapter **5**

Creating Tables and Listings

During the process of clinical trial analysis and reporting, most information is conveyed through the presentation of tables and listings. Generally, tables contain statistics about the data, while listings present the data in their raw form simply listed for review. These tables and listings can number in the hundreds for a new drug application down to just a dozen or so for an independent data monitoring committee (IDMC) report.

This chapter first explores the general approach to creating any statistical table or listing. Then it examines PROC REPORT and PROC TABULATE as possible stand-alone methods of clinical trial reporting. Next, examples of several common clinical trial tables are presented. Finally, issues concerning the appearance of the output are discussed.

Creating Tables

General Approach to Creating Tables

First it is worth taking the time to look at the conceptual framework behind creating statistical tables in SAS programming. Creating tables in a SAS program is a multiple-step process that remains independent of the SAS tools employed. These conceptual steps are as follows:

Step	Name	Description
1	Get data	This step involves pulling the data to be used into SAS. It often requires merging treatment or study population data with analysis data sets or some other data to be summarized/listed.
2	Manipulate data	On occasion the data being pulled into SAS for summarization and presentation are not ready for that purpose. In such cases, you may need to manipulate or create additional variables within the SAS program. Keep in mind that it is almost always better to create derived variables prior to this step in analysis data sets programming.
3	Create table statistics	This step involves calculating statistics for the tables that require summary or inferential statistics. Chapter 7 covers this step in greater detail.
4	Present output	This step involves presentation of the summarized data. There is a wealth of options available here for the statistical programmer.

Note that for generating tables there are single SAS procedures such as PROC TABULATE and PROC REPORT that can perform all of the preceding four steps. We will examine PROC TABULATE and PROC REPORT as possible reporting methods after you see what a typical clinical trial table looks like.

A Typical Clinical Trial Table

Clinical trial tables have certain common features. A sample annotated demographics table follows, with annotated descriptions of the key features.

```
Company/Trial Name                                              Page X of N
                                                                    ❶
                                   Table X
                     Demographics and Baseline Characteristics

         ----------------------------------------------------------------------
                         Active         Placebo    ❷      Overall
         Variable        (N=31)         (N=29)            (N=60)      P-value
         ----------------------------------------------------------------------

         Age (years)                                                 0.9528    ❸
            N            31             29                60
    ❹      Mean          51.4           50.1              50.8
            Standard Deviation  13.2    13.2              13.1
            Minimum      32             23                23
            Maximum      75             77                77

         Gender                                                      0.2681
            Male     ❺  22 ( 71.0%)    16 ( 55.2%)       38 ( 63.3%)
            Female       9 ( 29.0%)    12 ( 41.4%)       21 ( 35.0%)

         Race
            Black        18 ( 58.1%)   18 ( 62.1%)       36 ( 60.0%)  0.9270
            White        10 ( 27.6%)    8 ( 32.3%)       18 ( 30.0%)
            Other*        3 ( 10.3%)    3 (  9.7%)        6 ( 10.0%)

         ----------------------------------------------------------------------
      * Other includes Asian, Native American, and other races.  ❻
        Created by program_name.sas on mm/dd/yyyy.
```

Notes for the table:

❶ This is not a simple page counter. It also includes the total number of pages for the given piece of output.

❷ In the column heading you see "N=," which represents the number of subjects in the given column population.

❸ Here you have the *p*-values from inferential statistical comparisons.

❹ Note that the continuous statistics are presented with a row-based orientation.

❺ Categorical frequency counts are presented in a single item as "count (percentage)." These percentages can be calculated and formatted in many different ways, but they are always clustered together as a single column item.

❻ The footnote contains the name of the program that created the output as well as the date it was created.

We now examine PROC REPORT and PROC TABULATE as possible stand-alone methods of creating this demographics table.

Using PROC TABULATE to Create Clinical Trial Tables

You can use PROC TABULATE to produce a summary statistics matrix with very little effort. Here are the annotated demographics summary program, the annotated output and notes for the program, and a follow-up discussion of PROC TABULATE's capabilities.

Program 5.1 Using PROC TABULATE to Create a Summary of Demographics

```
**** INPUT SAMPLE DEMOGRAPHICS DATA;
data demog;
label subjid   = "Subject Number"
      trt      = "Treatment"
      gender   = "Gender"
      race     = "Race"
      age      = "Age";
input subjid trt gender race age @@;
datalines;
101 0 1 3 37   301 0 1 1 70   501 0 1 2 33   601 0 1 1 50   701 1 1 1 60
102 1 2 1 65   302 0 1 2 55   502 1 2 1 44   602 0 2 2 30   702 0 1 1 28
103 1 1 2 32   303 1 1 1 65   503 1 1 1 64   603 1 2 1 33   703 1 1 2 44
104 0 2 1 23   304 0 1 1 45   504 0 1 3 56   604 0 1 1 65   704 0 2 1 66
105 1 1 3 44   305 1 1 1 36   505 1 1 2 73   605 1 2 1 57   705 1 1 2 46
106 0 2 1 49   306 0 1 2 46   506 0 1 1 46   606 0 1 2 56   706 1 1 1 75
201 1 1 3 35   401 1 2 1 44   507 1 1 2 44   607 1 1 1 67   707 1 1 1 46
202 0 2 1 50   402 0 2 2 77   508 0 2 1 53   608 0 2 2 46   708 0 2 1 55
203 1 1 2 49   403 1 1 1 45   509 0 1 1 45   609 1 2 1 72   709 0 2 2 57
204 0 2 1 60   404 1 1 1 59   510 0 1 3 65   610 0 1 1 29   710 0 1 1 63
205 1 1 3 39   405 0 2 1 49   511 1 2 2 43   611 1 2 1 65   711 1 1 2 61
206 1 2 1 67   406 1 1 2 33   512 1 1 1 39   612 1 1 2 46   712 0 . 1 49
;
```

```
**** DEFINE VARIABLE FORMATS NEEDED FOR TABLE;
proc format;
   value trt
       1 = "Active"
       0 = "Placebo";
   value gender
       1 = "Male"
       2 = "Female";
   value race
       1 = "White"
       2 = "Black"
       3 = "Other*";
run;
```

❶

```
**** DEFINE OPTIONS FOR ASCII TEXT OUTPUT;
options nodate ls = 80 ps = 38 formchar = "|----|+|---+=|-/\<>*";

**** CREATE SUMMARY OF DEMOGRAPHICS WITH PROC TABULATE;
proc tabulate
   data = demog
   missing;

   class trt gender race;
   var age;
   table age = 'Age' *
               (n = 'n' * f = 8.
                mean = 'Mean' * f = 5.1
                std = 'Standard Deviation' * f = 5.1
                min = 'Min' * f = 3. Max = 'Max' * f = 3.)
           gender = 'Gender' *
               (n='n' * f = 3. colpctn = '%' * f = 4.1)
           race = 'Race' *
               (n = 'n' * f = 3. colpctn = '%' * f = 4.1),

           (trt = "  ") (all = 'Overall');

   format trt trt. race race. gender gender.;

   title1 'Table 5.1';
   title2 'Demographics and Baseline Characteristics';
   footnote1 "* Other includes Asian, Native American, and other"
             " races.";
   footnote2 "Created by %sysfunc(getoption(sysin)) on"
             " &sysdate9..";
run;
```

Program 5.1 produces the following output.

```
                          Table 5.1                     1          ❷
                Demographics and Baseline Characteristics


       ----------------------------------------------
       |               |Placebo | Active |Overall |
       |---------------+--------+--------+--------|
       |Age    |n       |     29|     31|     60|
       |       |--------+--------+--------+--------|
       |       |Mean    |   50.1|   51.4|   50.8|
       |       |--------+--------+--------+--------|
       |       |Standard|        |        |        |
       |       |Deviation|   13.2|   13.2|   13.1|
       |       |--------+--------+--------+--------|
       |       |Min     |     23|     32|     23|
       |       |--------+--------+--------+--------|
       |       |Max     |     77|     75|     77|
       |-------+--------+--------+--------+--------|
       |Gender |        |        |        |        |
       |-------+--------|        |        |        |
       |.      |n       |      1|      .|      1|          ❸
       |       |--------+--------+--------+--------|
       |       |%       |    3.4|      .|    1.7|
       |-------+--------+--------+--------+--------|
       |Male   |n       |     16|     22|     38|
       |       |--------+--------+--------+--------|
       |       |%       |   55.2|   71.0|   63.3|
       |-------+--------+--------+--------+--------|
       |Female |n       |     12|      9|     21|
       |       |--------+--------+--------+--------|
       |       |%       |   41.4|   29.0|   35.0|
       ----------------------------------------------


(Continued)    ❹
        * Other includes Asian, Native American, and other races.
        Created by C:\tabulate_sample.sas on 01MAR2005.

<PAGE BREAK HERE...>
```

```
                        Table 5.1                        2
          Demographics and Baseline Characteristics


     ---------------------------------------------------
     |                   |Placebo | Active |Overall |
     |------------------+--------+--------+--------|
     |Race    |         |        |        |        |
     |--------+---------|        |        |        |
     |White   |n        |     18|     18|     36|
     |        |---------+--------+--------+--------|
     |        |%        |   62.1|   58.1|   60.0|
     |--------+---------+--------+--------+--------|
     |Black   |n        |      8|     10|     18|
     |        |---------+--------+--------+--------|
     |        |%        |   27.6|   32.3|   30.0|
     |--------+---------+--------+--------+--------|
     |Other*  |n        |      3|      3|      6|
     |        |---------+--------+--------+--------|
     |        |%        |   10.3|    9.7|   10.0|
     ---------------------------------------------------


     * Other includes Asian, Native American, and other races.
     Created by C:\tabulate_sample.sas on 01MAR2005.
```

Notes for Program 5.1:

❶   The FORMCHAR option is specified to get the simple ASCII text results to appear in a platform independent way. We discuss output formatting later in this chapter.

❷   Note that SAS provides a page counter, but not the "Page X of N" style page counter.

❸   There are several items about the body of the table to mention here. First, there is no "p-value" column, as PROC TABULATE generally produces only descriptive statistics. Second, the styles of the "n (%)" statistics are oriented with "n" and "%" in different rows when we wanted "n (%)" in the same row in the same "cell." Third, the field spanner for "Gender" is not indented differently from the summary statistic labels. Finally, note the missing (".") row category for gender. In the PROC TABULATE statement, if the MISSING option had been excluded, then the data for subject 712 would have been wrongly omitted from the output for all summarized variables, including age, gender, and race.

❹   In traditional ASCII text output, PROC TABULATE does provide a "continued" flag when the output spans multiple pages. (This is not the case with some ODS destinations, such as ODS RTF, where there is no "continued" flag.)

PROC TABULATE is an excellent tool for producing quick descriptive statistics on data, but it does not meet the typical needs of generating clinical trial tables, for several reasons:

1. It does not provide *p*-values beyond a simple *t*-test.

2. It does not present "n (%)" in a desired format.

3. Variable labels are not clearly differentiated from the summary statistics labels.

4. "Page X of N" pagination is not available.

5. Missing values are handled in an "all or nothing" way, so if you exclude missing values for one variable, you end up excluding all data for that record for the other variables in the summary.

6. It does not place population counts in the column heading.

Using PROC REPORT to Create Clinical Trial Tables

PROC REPORT is also capable of producing a summary of data with little effort. Here are the demographics table annotated SAS program, the annotated output and notes for the program, and a follow-up discussion of PROC REPORT's capabilities.

Program 5.2 Using PROC REPORT to Create a Summary of Demographics

```
****  INPUT SAMPLE DEMOGRAPHICS DATA;
data demog;
label subjid   = "Subject Number"
      trt      = "Treatment"
      gender   = "Gender"
      race     = "Race"
      age      = "Age";
input subjid trt gender race age @@;
datalines;
101 0 1 3 37   301 0 1 1 70   501 0 1 2 33   601 0 1 1 50   701 1 1 1 60
102 1 2 1 65   302 0 1 2 55   502 1 2 1 44   602 0 2 2 30   702 0 1 1 28
103 1 1 2 32   303 1 1 1 65   503 1 1 1 64   603 1 2 1 33   703 1 1 2 44
104 0 2 1 23   304 0 1 1 45   504 0 1 3 56   604 0 1 1 65   704 0 2 1 66
105 1 1 3 44   305 1 1 1 36   505 1 1 2 73   605 1 2 1 57   705 1 1 2 46
106 0 2 1 49   306 0 1 2 46   506 0 1 1 46   606 0 1 2 56   706 1 1 1 75
201 1 1 3 35   401 1 2 1 44   507 1 1 2 44   607 1 1 1 67   707 1 1 1 46
202 0 2 1 50   402 0 2 2 77   508 0 2 1 53   608 0 2 2 46   708 0 2 1 55
203 1 1 2 49   403 1 1 1 45   509 0 1 1 45   609 1 2 1 72   709 0 2 2 57
204 0 2 1 60   404 1 1 1 59   510 0 1 3 65   610 0 1 1 29   710 0 1 1 63
205 1 1 3 39   405 0 2 1 49   511 1 2 2 43   611 1 2 1 65   711 1 1 2 61
206 1 2 1 67   406 1 1 2 33   512 1 1 1 39   612 1 1 2 46   712 0 . 1 49
;
```

```
**** DEFINE VARIABLE FORMATS NEEDED FOR TABLE;
proc format;
   value trt
      1 = "-- Active --"
      0 = "-- Placebo --";
   value gender
      1 = "Male"
      2 = "Female";
   value race
      1 = "White"
      2 = "Black"
      3 = "Other*";
run;

**** DEFINE OPTIONS FOR ASCII TEXT OUTPUT;
options nodate nocenter ls = 70
        formchar = "|----|+|---+=|-/\<>*";

**** CREATE SUMMARY OF DEMOGRAPHICS WITH PROC TABULATE;
proc report
   data = demog
   nowindows
   missing                                                    ❶
   headline;

   column ('--' trt,
          ( ("-- Age --"
              age = agen age = agemean age = agestd age = agemin
              age = agemax)
              gender,(gender = gendern genderpct)
              race,(race = racen racepct)));             ❷

   define trt /across format = trt. "  ";
   define agen /analysis n format = 3. 'N';
   define agemean /analysis mean format = 5.3 'Mean';
   define agestd /analysis std format = 5.3 'SD';
   define agemin /analysis min format = 3. 'Min';
   define agemax /analysis max format = 3. 'Max';

   define gender /across "-- Gender --" format = gender.;
   define gendern /analysis n format = 3. 'N';
   define genderpct /computed format = percent5. '(%)';
   define race /across "-- Race --" format = race.;
   define racen /analysis n format = 3. width = 6 'N';
   define racepct /computed format = percent5. '(%)';

   compute before;
```

```
            totga = sum(_c6_,_c8_,_c10_);
            totgp = sum(_c23_,_c25_,_c27_);
            totra = sum(_c12_,_c14_,_c16_);
            totrp = sum(_c29_,_c31_,_c33_);
            endcomp;
            compute genderpct;
            _c7_ = _c6_ / totga;
            _c9_ = _c8_ / totga;
            _c11_ = _c10_ / totga;
            _c24_ = _c23_ / totgp;
            _c26_ = _c25_ / totgp;
            _c28_ = _c27_ / totgp;
            endcomp;
            compute racepct;
            _c13_ = _c12_ / totra;
            _c15_ = _c14_ / totra;
            _c17_ = _c16_ / totra;
            _c30_ = _c29_ / totrp;
            _c32_ = _c31_ / totrp;
            _c34_ = _c33_ / totrp;
            endcomp;

            title1 'Table 5.2';
            title2 'Demographics and Baseline Characteristics';
            footnote1 "Created by %sysfunc(getoption(sysin))"
                      " on &sysdate9..";
        run;
```

❸

Program 5.2 produces the following output.

```
Table 5.2                                                              1
Demographics and Baseline Characteristics
----------------------------------------------------------------
                                                                ❹
------------------------ Active --------------------------
                            ------------ Gender -------------
----------- Age -----------       .       Female       Male
  N    Mean     SD  Min  Max    N    (%)    N    (%)    N    (%)
----------------------------------------------------------------
  31  51.35  13.23   32   75     .    .     9   29%    22   71%

Created by C:\report_sample.sas on 01MAR2005.


<PAGE BREAK HERE...>

Table 5.2                                                              2
Demographics and Baseline Characteristics

   ----------------------------------------------------------------

   ---------------- Active ----------------   ------ Placebo -------
   ---------------- Race ----------------
     Black          Other*         White      -------- Age ---------
     N    (%)        N    (%)       N    (%)    N    Mean      SD   Min
   ----------------------------------------------------------------
     10   32%        3   10%       18   58%    29  50.10   13.22    23

   Created by C:\report_sample.sas on 01MAR2005.


<PAGE BREAK HERE...>
```

(continued)

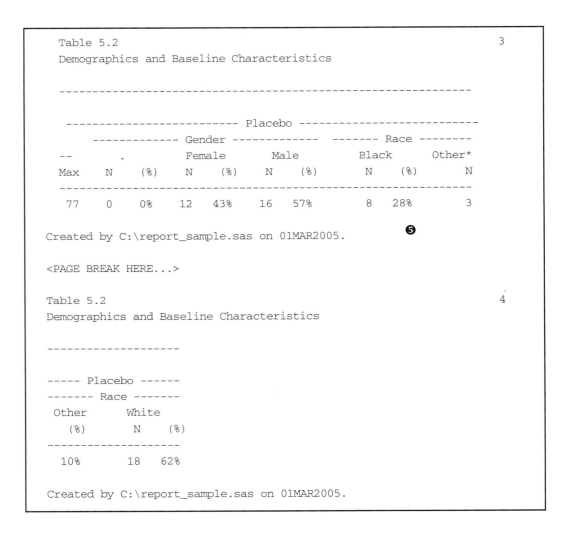

```
Table 5.2                                                          3
Demographics and Baseline Characteristics


   ------------------------------------------------------------


   ------------------------ Placebo -------------------------
      ------------ Gender -------------   ------- Race --------
   --       .         Female       Male       Black        Other*
   Max   N    (%)    N    (%)    N    (%)    N    (%)        N

   ------------------------------------------------------------
   77    0    0%    12   43%    16   57%     8   28%         3

Created by C:\report_sample.sas on 01MAR2005.            ❺


<PAGE BREAK HERE...>


Table 5.2                                                          4
Demographics and Baseline Characteristics


   --------------------


   ----- Placebo ------
   ------- Race -------
    Other       White
     (%)       N    (%)
   --------------------
    10%        18   62%


Created by C:\report_sample.sas on 01MAR2005.
```

Notes for Program 5.2:

❶ In the PROC REPORT statement, if the MISSING option had been excluded, then the data for subject 712 would have been wrongfully omitted from the output for all summarized variables, including age, gender, and race.

❷ Here you see the column definitions for **PROC REPORT**. Treatment, Gender, and Race are defined as ACROSS variables in order to get their categorical values to span columns as wanted.

❸ Note here the large number of COMPUTE block calculations that are needed in order to calculate the percentages for gender and race.

❹ Here you see that PROC REPORT is a column-based reporting tool. Instead of having the statistics flow down the page, PROC REPORT stretches the statistics across the page and onto several subsequent pages.

❺ As you see from note 1, specifying the MISSING option is essential. However, in this case even though there is a missing gender value, PROC REPORT shows that 0 subjects are missing gender. By design PROC REPORT places only non-missing values in the N statistic even if the category being summarized is in fact the missing category. In this case, the N for the patients missing gender should be "1."

PROC REPORT as a stand-alone tool is not very useful for creating clinical trial tables, for several reasons:

1. It produces only columnar-style reports and not the matrix-style reports required by clinical trial reporting.

2. It does not provide *p*-values beyond a simple *t*-test.

3. "Page X of N" pagination is not available.

4. Missing values are handled in an "all or nothing" way, so if you exclude missing values for one variable, you end up excluding all data for that record for the other variables in the summary. Even if missing values are included with the MISSING option, the N statistic is not calculated as you might expect for missing variables.

5. It does not place population counts in the column heading.

If PROC REPORT and PROC TABULATE by themselves cannot produce the type of tables desired, then alternative means of generating clinical trial tables must be explored.

Creating Continuous/Categorical Summary Tables

Much like the previous demographics table example, the majority of the tables generated for clinical trial reporting compare a set of continuous and/or categorical variables across treatment groups. The purpose is to see whether therapy groups are comparable or whether they differ in some way. These kinds of tables include the following:

- Patient disposition tables, where you want to see the flow of patients through a trial. Disposition tables show you the number of patients who were randomized, treated, and discontinued (and why) or who completed the study, and other protocol compliance counts.

- Demographics/baseline characteristics tables, where you want to ascertain whether the population under study is comparable between therapy groups at baseline.

- Efficacy results tables, where you want to see which of the therapies treats the underlying disease state better.

- Safety results tables, which contain adverse event, laboratory, and vital signs data summaries.

Now let's revisit the sample demographics table created previously, as it contains both categorical and continuous statistics in a single table. The following example relies on DATA step programming, SAS macro variables, some statistical procedures, and a final PROC REPORT for table presentation. Here is the demographics table annotated SAS program, followed by program notes and the program output.

Program 5.3 Creating a Typical Summary of Demographics

```
**** INPUT SAMPLE DEMOGRAPHICS DATA.;
data demog;
label subjid  = "Subject Number"
      trt     = "Treatment"
      gender  = "Gender"
      race    = "Race"
      age     = "Age";
input subjid trt gender race age @@;
datalines;
101 0 1 3 37   301 0 1 1 70   501 0 1 2 33   601 0 1 1 50   701 1 1 1 60
102 1 2 1 65   302 0 1 2 55   502 1 2 1 44   602 0 2 2 30   702 0 1 1 28
103 1 1 2 32   303 1 1 1 65   503 1 1 1 64   603 1 2 1 33   703 1 1 2 44
104 0 2 1 23   304 0 1 1 45   504 0 1 3 56   604 0 1 1 65   704 0 2 1 66
105 1 1 3 44   305 1 1 1 36   505 1 1 2 73   605 1 2 1 57   705 1 1 2 46
106 0 2 1 49   306 0 1 2 46   506 0 1 1 46   606 0 1 2 56   706 1 1 1 75
201 1 1 3 35   401 1 2 1 44   507 1 1 2 44   607 1 1 1 67   707 1 1 1 46
202 0 2 1 50   402 0 2 2 77   508 0 2 1 53   608 0 2 2 46   708 0 2 1 55
203 1 1 2 49   403 1 1 1 45   509 0 1 1 45   609 1 2 1 72   709 0 2 2 57
204 0 2 1 60   404 1 1 1 59   510 0 1 3 65   610 0 1 1 29   710 0 1 1 63
205 1 1 3 39   405 0 2 1 49   511 1 2 2 43   611 1 2 1 65   711 1 1 2 61
206 1 2 1 67   406 1 1 2 33   512 1 1 1 39   612 1 1 2 46   712 0 . 1 49
;
run;

**** CREATE FORMATS NEEDED FOR TABLE ROWS.;
proc format;
   value gender
      1 = "Male"
      2 = "Female";
```

```
      value race
         1 = "White"
         2 = "Black"
         3 = "Other*";
run;


**** DUPLICATE THE INCOMING DATA SET FOR OVERALL COLUMN                    ❶
**** CALCULATIONS SO NOW TRT HAS VALUES 0 = PLACEBO, 1 = ACTIVE,
**** AND 2 = OVERALL.;
data demog;
   set demog;
   output;
   trt = 2;
   output;
run;


**** AGE STATISTICS PROGRAMMING ********************************;
**** GET P VALUE FROM NON PARAMETRIC COMPARISON OF AGE MEANS.;
proc npar1way                                                              ❷
   data = demog
   wilcoxon
   noprint;
      where trt in (0,1);
      class trt;
      var age;
      output out = pvalue wilcoxon;
run;


proc sort
   data = demog;
      by trt;
run;


***** GET AGE DESCRIPTIVE STATISTICS N, MEAN, STD, MIN, AND MAX.;
proc univariate
   data = demog noprint;
      by trt;
      var age;
      output out = age
               n = _n mean = _mean std = _std min = _min
               max = _max;
run;


**** FORMAT AGE DESCRIPTIVE STATISTICS FOR THE TABLE.;
data age;
   set age;
```

```
   format n mean sd min max $14.;
   drop _n _mean _std _min _max;

   n = put(_n,3.);
   mean = put(_mean,7.1);
   std = put(_std,8.2);
   min = put(_min,7.1);
   max = put(_max,7.1);
run;

**** TRANSPOSE AGE DESCRIPTIVE STATISTICS INTO COLUMNS.;
proc transpose
   data = age
   out = age
   prefix = col;
      var n mean std min max;
      id trt;
run;

**** CREATE AGE FIRST ROW FOR THE TABLE.;
data label;
   set pvalue(keep = p2_wil rename = (p2_wil = pvalue));
   length label $ 85;
   label = "Age (years)";
run;

**** APPEND AGE DESCRIPTIVE STATISTICS TO AGE P VALUE ROW AND
**** CREATE AGE DESCRIPTIVE STATISTIC ROW LABELS.;
data age;
   length label $ 85 col0 col1 col2 $ 25 ;
   set label age;

   keep label col0 col1 col2 pvalue ;
   if _n_ > 1 then
      select;
         when(_NAME_ = 'n')    label = "      N";
         when(_NAME_ = 'mean') label = "      Mean";
         when(_NAME_ = 'std')  label = "      Standard Deviation";
         when(_NAME_ = 'min')  label = "      Minimum";
         when(_NAME_ = 'max')  label = "      Maximum";
         otherwise;
      end;
run;
**** END OF AGE STATISTICS PROGRAMMING ************************;
```

```
**** GENDER STATISTICS PROGRAMMING *****************************;
**** GET SIMPLE FREQUENCY COUNTS FOR GENDER.;
proc freq
   data = demog
   noprint;
      where trt ne .;
      tables trt * gender / missing outpct out = gender;
run;

**** FORMAT GENDER N(%) AS DESIRED.;
data gender;
   set gender;
      where gender ne .;
      length value $25;
      value = put(count,4.) || ' (' || put(pct_row,5.1)||'%)';
run;

proc sort
   data = gender;
      by gender;
run;

**** TRANSPOSE THE GENDER SUMMARY STATISTICS.;
proc transpose
   data = gender
   out = gender(drop = _name_)
   prefix = col;
      by gender;
      var value;
      id trt;
run;

**** PERFORM CHI-SQUARE ON GENDER COMPARING ACTIVE VS PLACEBO.;
proc freq
   data = demog
   noprint;
      where gender ne . and trt not in (.,2);
      table gender * trt / chisq;
      output out = pvalue pchi;
run;

**** CREATE GENDER FIRST ROW FOR THE TABLE.;
data label;
   set pvalue(keep = p_pchi rename = (p_pchi = pvalue));
   length label $ 85;
   label = "Gender";
run;
```

```
**** APPEND GENDER DESCRIPTIVE STATISTICS TO GENDER P VALUE ROW
**** AND CREATE GENDER DESCRIPTIVE STATISTIC ROW LABELS.;
data gender;
   length label $ 85 col0 col1 col2 $ 25 ;
   set label gender;

   keep label col0 col1 col2 pvalue ;
   if _n_ > 1 then
       label= "      " || put(gender,gender.);
run;
**** END OF GENDER STATISTICS PROGRAMMING *********************;


**** RACE STATISTICS PROGRAMMING ****************************;
**** GET SIMPLE FREQUENCY COUNTS FOR RACE;
proc freq
   data = demog
   noprint;
       where trt ne .;
       tables trt * race / missing outpct out = race;
run;

**** FORMAT RACE N(%) AS DESIRED;
data race;
   set race;
       where race ne .;
       length value $25;
       value = put(count,4.) || ' (' || put(pct_row,5.1)||'%)';
run;

proc sort
   data = race;
       by race;
run;

**** TRANSPOSE THE RACE SUMMARY STATISTICS;
proc transpose
   data = race
   out = race(drop = _name_)
   prefix=col;
       by race;
       var value;
       id trt;
run;
```

```
**** PERFORM FISHER'S EXACT ON RACE COMPARING ACTIVE VS PLACEBO.;
proc freq
   data = demog
   noprint;
      where race ne . and trt not in (.,2);
      table race * trt / exact;
      output out = pvalue exact;
run;

**** CREATE RACE FIRST ROW FOR THE TABLE.;
data label;
   set pvalue(keep = xp2_fish rename = (xp2_fish = pvalue));
   length label $ 85;
   label = "Race";
run;

**** APPEND RACE DESCRIPTIVE STATISTICS TO RACE P VALUE ROW AND
**** CREATE RACE DESCRIPTIVE STATISTIC ROW LABELS.;
data race;
   length label $ 85 col0 col1 col2 $ 25 ;
   set label race;

   keep label col0 col1 col2 pvalue ;
   if _n_ > 1 then
      label= "       " || put(race,race.);
run;
**** END OF RACE STATISTICS PROGRAMMING ***********************;


**** CONCATENATE AGE, GENDER, AND RACE STATISTICS AND CREATE
**** GROUPING GROUP VARIABLE FOR LINE SKIPPING IN PROC REPORT.;
data forreport;
   set age(in = in1)
       gender(in = in2)
       race(in = in3);

       group = sum(in1 * 1, in2 * 2, in3 * 3);
run;


**** DEFINE THREE MACRO VARIABLES &N0, &N1, AND &NT THAT ARE USED
**** IN THE COLUMN HEADERS FOR "PLACEBO," "ACTIVE" AND "OVERALL"
**** THERAPY GROUPS.;
data _null_;
   set demog end = eof;
```

❸

❹

```
            **** CREATE COUNTER FOR N0 = PLACEBO, N1 = ACTIVE.;
            if trt = 0 then
                n0 + 1;
            else if trt = 1 then
                n1 + 1;

            **** CREATE OVERALL COUNTER NT.;
            nt + 1;

            **** CREATE MACRO VARIABLES &N0, &N1, AND &NT.;
            if eof then
                do;
                    call symput("n0",compress('(N='||put(n0,4.) || ')'));
                    call symput("n1",compress('(N='||put(n1,4.) || ')'));
                    call symput("nt",compress('(N='||put(nt,4.) || ')'));
                end;
run;

**** USE PROC REPORT TO WRITE THE TABLE TO FILE.;
options nonumber nodate ls = 84 missing = " "
        formchar = "|----|+|---+=|-/\<>*";

proc report
    data = forreport
    nowindows
    spacing = 1
    headline
    headskip
    split = "|";

    columns ("--" group label col1 col0 col2 pvalue);

    define group    /order order = internal noprint;
    define label    /display width=23 " ";
    define col0     /display center width = 14 "Placebo|&n0";
    define col1     /display center width = 14 "Active|&n1";
    define col2     /display center width = 14 "Overall|&nt";
    define pvalue   /display center width = 14 " |P-value**"
                    f = pvalue6.4;

    break after group / skip;

    title1 "Company
                  "                          ";
    title2 "Protocol Name
                  "                          ";
```

❺

```
      title3 "Table 5.3";
      title4 "Demographics";

      footnote1 "-------------------------------------------"
                "-------------------------------------------";
      footnote2 "*  Other includes Asian, Native Amerian, and other"
                " races.                                    ";
      footnote3 "** P-values:  Age = Wilcoxon rank-sum, Gender "
                "= Pearson's chi-square,                    ";
      footnote4 "                   Race = Fisher's exact test. "
                "                                           ";
      footnote5 "Created by %sysfunc(getoption(sysin)) on"
                " &sysdate9..";
   run;
```

Notes for Program 5.3:

❶ This program takes the approach of doubling the data to be summarized and assigning those data to trt = 2. By doubling the data, it is easier to treat the "Overall" column as just another treatment group in the descriptive statistics. This is fine so long as trt = 2 is dropped from any inferential statistics.

❷ Because "age" is not normally distributed here, the Wilcoxon signed rank test is used to calculate the *p*-value and is placed into a data set called "pvalue." (Inferential statistics are discussed further in Chapter 7.)

❸ "Race" is handled precisely as "gender" was, with the exception that Fisher's exact test is used for the therapy comparison between "Active" and "Placebo."

❹ Macro variables are used here to put values into the header of the PROC REPORT. Note that these macro variables could have been used for denominators for percentage calculations as well.

❺ Because simple ASCII text is the destination for the output, some TITLE and FOOTNOTE statements had to be padded on the right with spaces to get titles and footnotes to left align.

Program 5.3 produces the following output.

```
Company
Protocol Name
                                    Table 5.3
                                  Demographics

        ----------------------------------------------------------------------
                            Active        Placebo       Overall
                            (N=31)        (N=29)        (N=120)      P-value**
        ----------------------------------------------------------------------

Age (years)                                                          0.9528
        N                     31            29            60
        Mean                 51.4          50.1          50.8
        Standard Deviation   13.23         13.22         13.13
        Minimum              32.0          23.0          23.0
        Maximum              75.0          77.0          77.0

Gender                                                               0.2681
        Male             22 ( 71.0%)   16 ( 55.2%)   38 ( 63.3%)
        Female            9 ( 29.0%)   12 ( 41.4%)   21 ( 35.0%)

Race                                                                 0.9270
        White            18 ( 58.1%)   18 ( 62.1%)   36 ( 60.0%)
        Black            10 ( 32.3%)    8 ( 27.6%)   18 ( 30.0%)
        Other*            3 (  9.7%)    3 ( 10.3%)    6 ( 10.0%)


        ----------------------------------------------------------------------
 *   Other includes Asian, Native Amerian, and other races.
 **  P-values:  Age = Wilcoxon rank-sum, Gender = Pearson's chi-square,
                Race = Fisher's exact test.
             Created by C:\t_demog.sas on 12MAR2005.
```

This table has everything you would expect in a typical clinical trial summary of demographics, with the exception of the "Page X of N" pagination. You will see how to do that later in this chapter, when appearance options are discussed.

Creating Adverse Event Summaries

There are myriad ways in which adverse event data can be summarized. Adverse events are summarized by overall occurrence, by maximum severity, and by maximum relatedness to drug, and may be subset by many different subpopulations, including patients who died during the trial or discontinued treatment because of adverse events. Here we dissect a common adverse event summary—the summary of events by maximum severity.

The following is a table specification, or "*table shell*," for the summary of adverse events by body system, preferred term, and maximum severity. As a rule for this summary, a patient should be counted only once at maximum severity within each subgrouping. Denominators should be calculated as the sum of all patients who had the given treatment in the demographics file.

```
                              Table 5.4
                           Adverse Events
             By Body System, Preferred Term, and Greatest Severity


        ------------------------------------------------------------
        Body System            Active         Placebo        Overall
            Preferred Term      (N=x)          (N=x)          (N=x)
        ------------------------------------------------------------
        Any Event              xx (xxx.x%)    xx (xxx.x%)    xx (xxx.x%)
                Mild           xx (xxx.x%)    xx (xxx.x%)    xx (xxx.x%)
                Moderate       xx (xxx.x%)    xx (xxx.x%)    xx (xxx.x%)
                Severe         xx (xxx.x%)    xx (xxx.x%)    xx (xxx.x%)


        System Organ Class     xx (xxx.x%)    xx (xxx.x%)    xx (xxx.x%)
                Mild           xx (xxx.x%)    xx (xxx.x%)    xx (xxx.x%)
                Moderate       xx (xxx.x%)    xx (xxx.x%)    xx (xxx.x%)
                Severe         xx (xxx.x%)    xx (xxx.x%)    xx (xxx.x%)


            Preferred Term     xx (xxx.x%)    xx (xxx.x%)    xx (xxx.x%)
                Mild           xx (xxx.x%)    xx (xxx.x%)    xx (xxx.x%)
                Moderate       xx (xxx.x%)    xx (xxx.x%)    xx (xxx.x%)
                Severe         xx (xxx.x%)    xx (xxx.x%)    xx (xxx.x%)
```

(continued)

(continued)

```
    Preferred Term       xx (xxx.x%)     xx (xxx.x%)     xx (xxx.x%)
           Mild          xx (xxx.x%)     xx (xxx.x%)     xx (xxx.x%)
           Moderate      xx (xxx.x%)     xx (xxx.x%)     xx (xxx.x%)
           Severe        xx (xxx.x%)     xx (xxx.x%)     xx (xxx.x%)


  etc.
```

The following example relies on DATA step programming, a few SAS macro variables, and a final DATA _NULL_ step with PUT statements for custom table presentation. Here are the adverse event summary annotated SAS program, notes for the program, and the output.

Program 5.4 Summary of Adverse Events by Maximum Severity

```
**** INPUT SAMPLE TREATMENT DATA.;
data treat;
label subjid       = "Subject Number"
      trtcd        = "Treatment";
input subjid trtcd @@;
datalines;
101 1   102 0   103 0   104 1   105 0   106 0   107 1   108 1   109 0   110 1
111 0   112 0   113 0   114 1   115 0   116 1   117 0   118 1   119 1   120 1
121 1   122 0   123 1   124 0   125 1   126 1   127 0   128 1   129 1   130 1
131 1   132 0   133 1   134 0   135 1   136 1   137 0   138 1   139 1   140 1
141 1   142 0   143 1   144 0   145 1   146 1   147 0   148 1   149 1   150 1
151 1   152 0   153 1   154 0   155 1   156 1   157 0   158 1   159 1   160 1
161 1   162 0   163 1   164 0   165 1   166 1   167 0   168 1   169 1   170 1
;
run;


**** INPUT SAMPLE ADVERSE EVENT DATA.;
data ae;
label subjid   = "Subject Number"
      aebodsys = "Body System of Event"
      aedecod  = "Preferred Term for Event"
      aerel    = "Relatedness: 1=not,2=possibly,3=probably"
      aesev    = "Severity/Intensity:1=mild,2=moderate,3=severe";
input subjid 1-3 aerel 5 aesev 7
      aebodsys $ 9-34 aedecod $ 38-62;
datalines;
101 1 1 Cardiac disorders          Atrial flutter
101 2 1 Gastrointestinal disorders Constipation
```

```
102 2 2 Cardiac disorders           Cardiac failure
102 1 1 Psychiatric disorders       Delirium
103 1 1 Cardiac disorders           Palpitations
103 1 2 Cardiac disorders           Palpitations
103 2 2 Cardiac disorders           Tachycardia
115 3 2 Gastrointestinal disorders  Abdominal pain
115 3 1 Gastrointestinal disorders  Anal ulcer
116 2 1 Gastrointestinal disorders  Constipation
117 2 2 Gastrointestinal disorders  Dyspepsia
118 3 3 Gastrointestinal disorders  Flatulence
119 1 3 Gastrointestinal disorders  Hiatus hernia
130 1 1 Nervous system disorders    Convulsion
131 2 2 Nervous system disorders    Dizziness
132 1 1 Nervous system disorders    Essential tremor
135 1 3 Psychiatric disorders       Confusional state
140 1 1 Psychiatric disorders       Delirium
140 2 1 Psychiatric disorders       Sleep disorder
141 1 3 Cardiac disorders           Palpitations
;
run;

**** CREATE FORMAT FOR AE SEVERITY.;
proc format;
   value aesev
      1 = "Mild"
      2 = "Moderate"
      3 = "Severe";
run;

**** PERFORM A SIMPLE COUNT OF EACH TREATMENT ARM AND OUTPUT.
**** RESULT AS MACRO VARIABLES.  N1 = 1ST COLUMN N FOR ACTIVE
**** THERAPY, N2 = 2ND COLUMN N FOR PLACEBO, N3 REPRESENTS THE
**** 3RD COLUMN TOTAL N.;
data _null_;
   set treat end = eof;

   **** INCREMENT (AND RETAIN) EACH TREATMENT COUNTER.;
   if trtcd = 1 then
      n1 + 1;
   else if trtcd = 0 then
      n2 + 1;

   **** INCREMENT (AND RETAIN) TOTAL COUNTER.;
   n3 + 1;

   **** AT THE END OF THE FILE, CREATE &N1, &N2, AND &N3.;
   if eof then
```

❶

```
      do;
         call symput("n1", put(n1,4.));
         call symput("n2", put(n2,4.));
         call symput("n3", put(n3,4.));
      end;
run;

proc sort
   data = ae;
      by subjid;
run;

proc sort
   data = treat;
      by subjid;
run;

***** MERGE ADVERSE EVENT AND DEMOGRAPHICS DATA;
data ae;
   merge treat(in = intreat) ae(in = inae);
      by subjid;

      if intreat and inae;
run;


**** CALCULATE ANY EVENT LEVEL COUNTS.  THIS IS THE FIRST ROW IN
**** THE SUMMARY.;
data anyevent;
   set ae end = eof;
      by subjid;

      keep rowlabel count1 count2 count3;

      **** KEEP ONLY LAST RECORD PER SUBJECT AS WE ONLY WANT TO
      **** COUNT A PATIENT ONCE IF THEY HAD ANY ADVERSE EVENTS.;
      if last.subjid;

      **** INCREMENT (AND RETAIN) EACH AE COUNT.;
      if trtcd = 1 then
         count1 + 1;
      else if trtcd = 0 then
         count2 + 1;

      **** INCREMENT (AND RETAIN) TOTAL AE COUNT.;
      count3 + 1;
```

```
          **** KEEP LAST RECORD OF THE FILE WITH TOTALS.;
          if eof;

          **** CREATE ROW LABEL FOR REPORT.;
          length rowlabel $ 30;
          rowlabel = "Any Event";
run;


**** CALCULATE ANY EVENT BY MAXIMUM SEVERITY LEVEL COUNTS.  THIS
**** IS THE BY SEVERITY BREAKDOWN UNDER THE FIRST ROW OF THE
**** SUMMARY.;
proc sort
   data = ae
   out = bysev;
      by subjid aesev;

**** KEEP ONLY LAST RECORD PER SUBJECT AT HIGHEST SEVERITY AS WE
**** ONLY WANT TO COUNT A PATIENT ONCE AT MAX SEVERITY IF THEY
**** HAD ANY ADVERSE EVENTS.;
data bysev;
   set bysev;
      by subjid aesev;

      if last.subjid;
run;

proc sort
   data = bysev;
      by aesev;
run;

data bysev;
   set bysev end = eof;
      by aesev;

      keep rowlabel count1 count2 count3;

      **** INITIALIZE AE COUNTERS TO ZERO AT EACH SEVERITY LEVEL.;
      if first.aesev then
         do;
            count1 = 0;
            count2 = 0;
            count3 = 0;
         end;

      **** INCREMENT (AND RETAIN) EACH AE COUNT.;
```

```
            if trtcd = 1 then
                count1 + 1;
            else if trtcd = 0 then
                count2 + 1;

            **** INCREMENT (AND RETAIN) TOTAL COUNT.;
            count3 + 1;

            **** KEEP LAST RECORD WITHIN EACH SEVERITY LEVEL.;
            if last.aesev;

            **** CREATE ROW LABEL FOR REPORT.;
            length rowlabel $ 30;
            rowlabel = "        " || put(aesev, aesev.);
run;


**** CALCULATE BODY SYSTEM BY MAXIMUM SEVERITY LEVEL COUNTS.
**** THIS IS THE BY SEVERITY BREAKDOWN UNDER THE BODY SYSTEMS OF
**** THE SUMMARY.;
proc sort
    data = ae
    out = bysys_sev;
        by subjid aebodsys aesev;
run;

**** KEEP ONLY LAST RECORD PER SUBJECT PER BODY SYSTEM AT HIGHEST
**** SEVERITY AS WE ONLY WANT TO COUNT A PATIENT ONCE AT MAX
**** SEVERITY WITHIN A BODY SYSTEM.;
data bysys_sev;
    set bysys_sev;
        by subjid aebodsys aesev;

        if last.aebodsys;
run;

proc sort
    data = bysys_sev;
        by aebodsys aesev;
run;

data bysys_sev;
    set bysys_sev;
        by aebodsys aesev;

        keep aebodsys rowlabel count1 count2 count3;
```

```
     **** INITIALIZE COUNTERS TO ZERO AT EACH SEVERITY LEVEL.;
     if first.aesev then
        do;
           count1 = 0;
           count2 = 0;
           count3 = 0;
        end;

     **** INCREMENT (AND RETAIN) EACH AE COUNT.;
     if trtcd = 1 then
        count1 + 1;
     else if trtcd = 0 then
        count2 + 1;

     **** INCREMENT (AND RETAIN) TOTAL COUNT.;
     count3 + 1;

     **** KEEP LAST RECORD FOR EACH BODY SYSTEM SEVERITY LEVEL.;
     if last.aesev;

     **** CREATE ROW LABEL FOR REPORT.;
     length rowlabel $ 30;
     rowlabel = "       " || put(aesev, aesev.);
run;



**** CALCULATE BODY SYSTEM LEVEL AE COUNTS.  THIS IS DONE BY
**** ADDING UP THE BODY SYSTEM BY SEVERITY COUNTS.;
data bysys;
   set bysys_sev(rename = (count1 = _count1
                           count2 = _count2
                           count3 = _count3));
      by aebodsys;

      keep aebodsys rowlabel count1 count2 count3;

      **** INITIALIZE COUNTERS TO ZERO AT EACH NEW BODY SYSTEM.;
      if first.aebodsys then
         do;
            count1 = 0;
            count2 = 0;
            count3 = 0;
         end;

      **** INCREMENT (AND RETAIN) EACH AE COUNT.;
      count1 + _count1;
```

```
         count2 + _count2;
         count3 + _count3;

         **** KEEP LAST RECORD WITHIN EACH BODY SYSTEM.;
         if last.aebodsys;

         **** CREATE ROW LABEL FOR REPORT.;
         length rowlabel $ 30;
         rowlabel = aebodsys;
run;


**** INTERLEAVE OVERALL BODY SYSTEM COUNTS WITH BY SEVERITY
**** COUNTS.;
data bysys;
   set bysys bysys_sev;
      by aebodsys;
run;


**** CALCULATE PREFERRED TERM BY MAXIMUM SEVERITY LEVEL COUNTS.
**** THIS IS THE BY SEVERITY BREAKDOWN UNDER THE PREFERRED TERMS
**** IN THE SUMMARY.;
proc sort
   data = ae
   out = byterm_sev;
      by subjid aebodsys aedecod aesev;
run;

**** KEEP ONLY LAST RECORD PER SUBJECT PER BODY SYSTEM PER
**** ADVERSE EVENT AT HIGHEST SEVERITY AS WE ONLY WANT TO COUNT A
**** PATIENT ONCE AT MAX SEVERITY WITHIN A PREFERRED TERM.;
data byterm_sev;
   set byterm_sev;
      by subjid aebodsys aedecod aesev;

      if last.aedecod;
run;

proc sort
   data = byterm_sev;
      by aebodsys aedecod aesev;
run;

data byterm_sev;
   set byterm_sev;
      by aebodsys aedecod aesev;
```

❷

```
      keep aebodsys aedecod rowlabel count1 count2 count3;

   **** INITIALIZE COUNTERS TO ZERO AT EACH SEVERITY LEVEL.;
   if first.aesev then
      do;
         count1 = 0;
         count2 = 0;
         count3 = 0;
      end;

   **** INCREMENT (AND RETAIN) EACH AE COUNT.;
   if trtcd = 1 then
      count1 + 1;
   else if trtcd = 0 then
      count2 + 1;

   **** INCREMENT (AND RETAIN) TOTAL COUNT.;
   count3 + 1;

   **** KEEP LAST RECORD FOR EACH PREF. TERM SEVERITY LEVEL.;
   if last.aesev;

   **** CREATE ROW LABEL FOR REPORT.;
   length rowlabel $ 30;
   rowlabel = "       " || put(aesev, aesev.) ;
run;

**** CALCULATE PREFERRED TERM LEVEL AE COUNTS.  THIS IS DONE BY
**** ADDING UP THE PREFERRED TERM BY SEVERITY COUNTS.;
data byterm;
   set byterm_sev(rename = (count1 = _count1
                            count2 = _count2
                            count3 = _count3));
      by aebodsys aedecod;

      keep aebodsys aedecod rowlabel count1 count2 count3;

   **** INITIALIZE COUNTERS TO ZERO AT EACH NEW PREF. TERM.;
   if first.aedecod then
      do;
         count1 = 0;
         count2 = 0;
         count3 = 0;
      end;
```

```
        **** INCREMENT (AND RETAIN) EACH AE COUNT.;
        count1 + _count1;
        count2 + _count2;
        count3 + _count3;

        **** KEEP LAST RECORD WITHIN EACH PREFERRED TERM.;
        if last.aedecod;

        **** CREATE ROW LABEL FOR REPORT.;
        length rowlabel $ 30;
        rowlabel = "    " || aedecod ;
run;
```

❸

```
**** INTERLEAVE PREFERRED TERM COUNTS WITH BY SEVERITY COUNTS.;
data byterm;
    set byterm byterm_sev;
        by aebodsys aedecod;
run;
```

❹

```
**** INTERLEAVE BODY SYSTEM COUNTS WITH PREFERRED TERM COUNTS.;
data bysys_byterm;
    set bysys byterm;
        by aebodsys;
run;
```

❺

```
**** SET ALL INTERMEDIATE DATA SETS TOGETHER AND CALCULATE
**** PERCENTAGES.;
data all;
    set anyevent
        bysev
        bysys_byterm;

        length col1 col2 col3 $ 10;

        **** CALCULATE %S AND CREATE N/% TEXT IN COL1-COL3.;
        if rowlabel ne '' then
            do;
                pct1 = (count1 / &n1) * 100;
                pct2 = (count2 / &n2) * 100;
                pct3 = (count3 / &n3) * 100;

                col1 = put(count1,3.) || " (" || put(pct1,3.) || "%)";
                col2 = put(count2,3.) || " (" || put(pct2,3.) || "%)";
                col3 = put(count3,3.) || " (" || put(pct3,3.) || "%)";
            end;
```

```
      **** CREATE SYSTEM_AND_TERM USED AS AN INDEX FOR INSERTING
      **** BLANK LINES AND PAGE BREAKS IN THE DATA _NULL_ BELOW.;
      length system_and_term $ 200;
      system_and_term = aebodsys || aedecod;
run;


**** WRITE AE SUMMARY TO FILE USING DATA _NULL_.;
options nodate nonumber;                                                ❻
title1 "Table 5.4";
title2 "Summary of Adverse Events";
title3 "By Body System, Preferred Term, and Greatest Severity";

data _null_;
   set all(sortedby = aebodsys system_and_term) end = eof;
      by aebodsys system_and_term;

      **** SET UP OUTPUT FILE OPTIONS.;
      file print titles linesleft = ll pagesize = 40 linesize = 70;

      **** DEFINE A NEW PAGE FLAG.  IF 1, THEN INSERT PAGE BREAK.;
      retain newpage 0;

      **** PRINT OUTPUT PAGE HEADER.;
      if _n_ = 1 or newpage = 1 then
         do;
            put @1 "-------------------------------"
                   "------------------------------------" /
                @1 "Body System" /
                @4 "Preferred Term" @33 "Active" @47 "Placebo"
                   @62 "Overall" /
                @7 "Severity" @33 "N=&n1" @48 "N=&n2" @63 "N=&n3" /
                @1 "------------------------------"
                   "------------------------------------" ;
            **** IF A BODY SYSTEM SPANS PAGES, REPEAT THE
            **** BODY SYSTEM WITH A CONTINUED INDICATOR.;
            if not first.aebodsys then
               put @1 aebodsys " (Continued)";
         end;

      **** PUT AE COUNTS AND PERCENTAGES ON THE PAGE.;
      put @1 rowlabel $40. @30 col1 $10. @45 col2 $10.
          @60 col3 $10.;

      **** RESET NEW PAGE FLAG.;
      newpage = 0;
```

```
                **** IF AT THE END OF THE PAGE, PUT A DOUBLE UNDERLINE.
                **** OTHERWISE IF AT THE END OF A PREFERRED TERM AND NEAR THE
                **** BOTTOM OF THE PAGE (LL <= 6) THEN PUT A PAGE BREAK.
                **** OTHERWISE IF AT THE END OF A PREFERRED TERM PUT A BLANK
                **** LINE.;
                if eof then
                   put @1 "-------------------------------"
                          "-------------------------------------" /
                       @1 "-------------------------------"
                          "-------------------------------------";
                else if last.system_and_term and ll <= 6 then
                   do;
                       put @1 "-------------------------------"
                              "-------------------------------------" /
                          @60 "(CONTINUED)";
                       put _page_;
                       newpage = 1;
                   end;
                else if last.system_and_term then
                   put;
        run;
```

Notes for Program 5.4:

❶ This DATA _NULL_ step defines three macro variables, &n1, &n2, and &n3, that are used in the column headings for "Placebo," "Active" and "Overall" therapy groups as well as for denominators to be used for percentage calculations. Note that the variable representing all patients who received therapy serves as the denominator, as any of them could possibly have had an adverse event.

❷ Here the counts for overall body system and body system at maximum severity are interleaved with a SET statement. This ensures that the summary flows properly, with the overall body system counts appearing above the severity counts.

❸ Here the counts for overall preferred term and preferred term at maximum severity are interleaved with a SET statement. This ensures that the summary flows properly, with the overall preferred term counts appearing above the severity counts.

❹ Here the body system counts are interleaved with the preferred term counts by body system. This ensures that the body system counts appear before the related preferred term counts in the output.

❺ Here the "Any Event" counts are placed before the body system and preferred term counts so that they appear first in the summary. At this point, the percentages are calculated by using &n0, &n1, and &nt as denominators. The columns (col1–col3) are created and formatted as "XXX (XXX%)." Finally, the "system_and_term" variable is created as an index to keep the data grouped by preferred term in the output.

❻ Here is the report written to file by using a DATA _NULL_ step. A DATA step with PUT statements is used here primarily because it can paginate between preferred term text and can provide continuation flags both at the end of a page and at the start of a new page for body systems that span pages. You almost always need intelligent page breaking for adverse event summaries, although this technique can be used for other summaries as well.

Program 5.4 produces the following output.

```
                              Table 5.4
                            Adverse Events
           By Body System, Preferred Term, and Greatest Severity
----------------------------------------------------------------------
Body System
      Preferred Term              Active         Placebo         Overall
           Severity               N=  45         N=  25          N=  70
----------------------------------------------------------------------
Any Event                      9 ( 20%)        5 ( 20%)       14 ( 20%)
         Mild                  4 (  9%)        1 (  4%)        5 (  7%)
         Moderate              1 (  2%)        4 ( 16%)        5 (  7%)
         Severe                4 (  9%)        0 (  0%)        4 (  6%)

Cardiac disorders              2 (  4%)        2 (  8%)        4 (  6%)
         Mild                  1 (  2%)        0 (  0%)        1 (  1%)
         Moderate              0 (  0%)        2 (  8%)        2 (  3%)
         Severe                1 (  2%)        0 (  0%)        1 (  1%)

   Atrial flutter              1 (  2%)        0 (  0%)        1 (  1%)
         Mild                  1 (  2%)        0 (  0%)        1 (  1%)

   Cardiac failure             0 (  0%)        1 (  4%)        1 (  1%)
         Moderate              0 (  0%)        1 (  4%)        1 (  1%)

   Palpitations                1 (  2%)        1 (  4%)        2 (  3%)
         Moderate              0 (  0%)        1 (  4%)        1 (  1%)
         Severe                1 (  2%)        0 (  0%)        1 (  1%)
```

(continued)

(continued)

```
     Tachycardia              0 (   0%)      1 (   4%)      1 (   1%)
       Moderate               0 (   0%)      1 (   4%)      1 (   1%)

Gastrointestinal disorders    4 (   9%)      2 (   8%)      6 (   9%)
       Mild                   2 (   4%)      0 (   0%)      2 (   3%)
       Moderate               0 (   0%)      2 (   8%)      2 (   3%)
       Severe                 2 (   4%)      0 (   0%)      2 (   3%)
--------------------------------------------------------------------
                                                           (CONTINUED)


<PAGE BREAK HERE...>


                            Table 5.4
                          Adverse Events
           By Body System, Preferred Term, and Greatest Severity
--------------------------------------------------------------------
Body System
   Preferred Term             Active        Placebo        Overall
      Severity               N=  45         N=  25         N=  70
--------------------------------------------------------------------
Gastrointestinal disorders  (Continued)
   Abdominal pain             0 (   0%)      1 (   4%)      1 (   1%)
      Moderate                0 (   0%)      1 (   4%)      1 (   1%)

   Anal ulcer                 0 (   0%)      1 (   4%)      1 (   1%)
      Mild                    0 (   0%)      1 (   4%)      1 (   1%)

   Constipation               2 (   4%)      0 (   0%)      2 (   3%)
      Mild                    2 (   4%)      0 (   0%)      2 (   3%)

   Dyspepsia                  0 (   0%)      1 (   4%)      1 (   1%)
      Moderate                0 (   0%)      1 (   4%)      1 (   1%)

   Flatulence                 1 (   2%)      0 (   0%)      1 (   1%)
      Severe                  1 (   2%)      0 (   0%)      1 (   1%)

   Hiatus hernia              1 (   2%)      0 (   0%)      1 (   1%)
      Severe                  1 (   2%)      0 (   0%)      1 (   1%)
```

(continued)

(continued)

```
Nervous system disorders        2 (   4%)      1 (   4%)      3 (   4%)
       Mild                     1 (   2%)      1 (   4%)      2 (   3%)
       Moderate                 1 (   2%)      0 (   0%)      1 (   1%)


   Convulsion                   1 (   2%)      0 (   0%)      1 (   1%)
       Mild                     1 (   2%)      0 (   0%)      1 (   1%)


   Dizziness                    1 (   2%)      0 (   0%)      1 (   1%)
       Moderate                 1 (   2%)      0 (   0%)      1 (   1%)
-----------------------------------------------------------------
                                                        (CONTINUED)

<PAGE BREAK HERE...>


                         Table 5.4
                      Adverse Events
       By Body System, Preferred Term, and Greatest Severity
-----------------------------------------------------------------
Body System
   Preferred Term              Active        Placebo        Overall
      Severity                 N=  45        N=  25         N=  70
-----------------------------------------------------------------
Nervous system disorders  (Continued)
   Essential tremor             0 (   0%)      1 (   4%)      1 (   1%)
       Mild                     0 (   0%)      1 (   4%)      1 (   1%)


Psychiatric disorders           2 (   4%)      1 (   4%)      3 (   4%)
       Mild                     1 (   2%)      1 (   4%)      2 (   3%)
       Severe                   1 (   2%)      0 (   0%)      1 (   1%)


   Confusional state            1 (   2%)      0 (   0%)      1 (   1%)
       Severe                   1 (   2%)      0 (   0%)      1 (   1%)


   Delirium                     1 (   2%)      1 (   4%)      2 (   3%)
       Mild                     1 (   2%)      1 (   4%)      2 (   3%)


   Sleep disorder               1 (   2%)      0 (   0%)      1 (   1%)
       Mild                     1 (   2%)      0 (   0%)      1 (   1%)
-----------------------------------------------------------------
-----------------------------------------------------------------
```

Note that by changing the "aesev" variable to the "aerel" variable throughout Program 5.4, you can easily change the previous adverse event summary to a summary of adverse events by maximum drug relatedness. Also, if you remove the maximum severity steps, you get a typical overall summary of adverse events by body system and preferred term. Since patient medical history data are also often coded with MedDRA, patient medical history data may be summarized much like an overall summary of adverse events. However, frequently medical histories are collected in a checklist/checkbox format so that using a coding dictionary is unnecessary.

Creating Concomitant or Prior Medication Tables

Concomitant or prior medication summaries are very similar to the adverse events summary you have just seen. The main difference is that most often medication tables do not have a high summary level, such as the body system found in adverse event summaries. When medication is coded with WHODrug, the ATC class can be used as the high-level data, but this is often not done because a given drug can belong to multiple ATC classes. However, you may be able to get your data management group to indicate the preferred ATC class when they are coding medications.

The following is a table specification for an overall summary of concomitant medications. As a rule for this summary, a patient should be counted only once per medication. Denominators are defined as the sum of all patients who received the specified study therapy.

```
                          Table 5.5
                 Summary of Concomitant Medication


      ---------------------------------------------------------------
                              Active        Placebo        Overall
      Preferred Medication Term  (N =x)        (N=x)          (N=x)
      ---------------------------------------------------------------
      Any Medication          xx (xxx.x%)   xx (xxx.x%)   xx (xxx.x%)


      Specific Medication #1   xx (xxx.x%)   xx (xxx.x%)   xx (xxx.x%)
      Specific Medication #2   xx (xxx.x%)   xx (xxx.x%)   xx (xxx.x%)
      Specific Medication #3   xx (xxx.x%)   xx (xxx.x%)   xx (xxx.x%)
      etc.
```

The following example relies on more PROC SQL, a little DATA step programming, a few SAS macro variables, and a final PROC REPORT for table presentation. Here are the concomitant medication summary annotated SAS program, notes for the program, and the output.

Program 5.5 Summary of Concomitant Medications

```
****  INPUT SAMPLE TREATMENT DATA.;
data treat;
label subjid      = "Subject Number"
      trtcd       = "Treatment";
input subjid trtcd @@;
datalines;
101 1   102 0   103 0   104 1   105 0   106 0   107 1   108 1   109 0   110 1
111 0   112 0   113 0   114 1   115 0   116 1   117 0   118 1   119 1   120 1
121 1   122 0   123 1   124 0   125 1   126 1   127 0   128 1   129 1   130 1
131 1   132 0   133 1   134 0   135 1   136 1   137 0   138 1   139 1   140 1
141 1   142 0   143 1   144 0   145 1   146 1   147 0   148 1   149 1   150 1
151 1   152 0   153 1   154 0   155 1   156 1   157 0   158 1   159 1   160 1
161 1   162 0   163 1   164 0   165 1   166 1   167 0   168 1   169 1   170 1
;

****  INPUT SAMPLE CONCOMITANT MEDICATION DATA.;
data cm;
label subjid      = "Subject Number"
      cmdecod     = "Standardized Medication Name";
input subjid 1-3 cmdecod $ 5-27;
datalines;
101 ACETYLSALICYLIC ACID
101 HYDROCORTISONE
102 VICODIN
102 POTASSIUM
102 IBUPROFEN
103 MAGNESIUM SULFATE
103 RINGER-LACTATE SOLUTION
115 LORAZEPAM
115 SODIUM BICARBONATE
116 POTASSIUM
117 MULTIVITAMIN
117 IBUPROFEN
119 IRON
130 FOLIC ACID
131 GABAPENTIN
132 DIPHENHYDRAMINE
135 SALMETEROL
140 HEPARIN
140 HEPARIN
140 NICOTINE
141 HYDROCORTISONE
141 IBUPROFEN
;
```

```
**** PERFORM A SIMPLE COUNT OF EACH TREATMENT ARM AND OUTPUT RESULT
**** AS MACRO VARIABLES.  N1 = 1ST COLUMN N FOR ACTIVE THERAPY,
**** N2 = 2ND COLUMN N FOR PLACEBO, N3 REPRESENTS THE 3RD COLUMN
**** TOTAL N.;
proc sql                                                              ❶
   noprint;

   **** PLACE THE NUMBER OF ACTIVE SUBJECTS IN &N1.;
   select count(distinct subjid) format = 3.
      into :n1
      from treat
      where trtcd = 1;
   **** PLACE THE NUMBER OF PLACEBO SUBJECTS IN &N2.;
   select count(distinct subjid) format = 3.
      into :n2
      from treat
      where trtcd = 0;
   **** PLACE THE TOTAL NUMBER OF SUBJECTS IN &N3.;
   select count(distinct subjid) format = 3.
      into :n3
      from treat
      where trtcd ne .;
quit;

***** MERGE CCONCOMITANT MEDICATIONS AND TREATMENT DATA.
***** KEEP RECORDS FOR SUBJECTS WHO HAD CONMEDS AND TOOK STUDY
***** THERAPY.  GET UNIQUE CONCOMITANT MEDICATIONS WITHIN PATIENTS.;
proc sql                                                              ❷
   noprint;
   create table cmtosum as
      select unique(c.cmdecod) as cmdecod, c.subjid, t.trtcd
         from cm as c, treat as t
         where c.subjid = t.subjid
         order by subjid, cmdecod;
quit;

**** TURN OFF LST OUTPUT.;
ods listing close;

**** GET MEDICATION COUNTS BY TREATMENT AND PLACE IN DATA SET
**** COUNTS.;                                                         ❸
ods output CrossTabFreqs = counts;
proc freq
   data = cmtosum;
      tables trtcd * cmdecod;
run;
```

```
ods output close;
ods listing;

proc sort
   data = counts;
      by cmdecod;
run;
```

❹

```
**** MERGE COUNTS DATA SET WITH ITSELF TO PUT THE THREE
**** TREATMENT COLUMNS SIDE BY SIDE FOR EACH CONMED.  CREATE GROUP
**** VARIABLE, WHICH IS USED TO CREATE BREAK LINE IN THE REPORT.
**** DEFINE COL1-COL3, WHICH ARE THE COUNT/% FORMATTED COLUMNS.;
data cm;
   merge counts(where = (trtcd = 1) rename = (frequency = count1))
         counts(where = (trtcd = 0) rename = (frequency = count2))
         counts(where = (trtcd = .) rename = (frequency = count3))
         end = eof;
      by cmdecod;

      keep cmdecod rowlabel col1-col3 group lastrec;
      length rowlabel $ 25 col1-col3 $ 10;

      **** LABEL "ANY MEDICATION" ROW AND PUT IN FIRST GROUP.
      **** BY MEDICATION COUNTS GO IN THE SECOND GROUP.;
      if cmdecod = '' then
         do;
            rowlabel = "ANY MEDICATION";
            group = 1;
         end;
      else
          do;
            rowlabel = cmdecod;
            group = 2;
          end;

      **** CALCULATE PERCENTAGES AND CREATE N/% TEXT IN COL1-COL3.;
      pct1 = (count1 / &n1) * 100;
      pct2 = (count2 / &n2) * 100;
      pct3 = (count3 / &n3) * 100;

      col1 = put(count1,3.) || " (" || put(pct1, 3.) || "%)";
      col2 = put(count2,3.) || " (" || put(pct2, 3.) || "%)";
      col3 = put(count3,3.) || " (" || put(pct3, 3.) || "%)";
```

```
                **** CREATE LASTERC FLAG AT THE END OF THE DATA SET FOR A
                **** SPECIAL END OF REPORT LINE IN PROC REPORT.;
                lastrec = eof;
         run;


         **** WRITE SUMMARY STATISTICS TO FILE USING PROC REPORT.;
         options   nonumber nodate;
         options formchar = "|----|+|---+=|-/\<>*";

         proc report
            data = cm
            nowindows
            spacing = 1
            headline
            ls = 80 ps = 20
            split = "|";

            columns ("--" lastrec group rowlabel col1 col2 col3);

            define lastrec  /display noprint;
            define group    /order noprint;
            define rowlabel /order width = 25 "Preferred Medication Term";
            define col1     /display center width = 14 "Active|N=&n1";
            define col2     /display center width = 14 "Placebo|N=&n2";
            define col3     /display center width = 14 "Total|N=&n3";

            break after group / skip;

            compute after _page_ / left;
            if not lastrec then
               contline = "(Continued)";
            else
               contline = "-----------";

            line @6 "----------------------------"
                    "----------------------------" contline $11.;
            endcomp;

         title1 "Table 5.5";
         title2 "Summary of Concomitant Medication";
         footnote1 "Created by %sysfunc(getoption(sysin)) on &sysdate9..";
         run;
```

❺

Notes for Program 5.5:

❶ Instead of using a DATA _NULL_ step as was done in Program 5.4, this program uses PROC SQL to define the three macro variables, &n1, &n2, and &n3, that are used in the column headings for "Active," "Placebo" and "Overall" therapy groups. The macro variables also serve as denominators to be used for percentage calculations.

❷ Here again PROC SQL is used where DATA steps may have been used before. In this single PROC SQL the treatment data are joined with the concomitant medications data and unique medications are selected within a patient.

❸ PROC FREQ is used to get summary counts here. In order to get the "Overall" counts, the "CrossTabFreqs" table output is produced using ODS, which gives totals as well as "Active" and "Placebo" counts. The totals are represented in this data set where "trtcd" is missing. If you use the /OUT= data set provided with the TABLES statement, you do not get totals. The ODS listing destination is closed and then reopened so that the subsequent PROC REPORT will produce output.

❹ This DATA step rearranges the "counts" data set created by PROC FREQ. The data set is essentially merged with itself three times in order to get each treatment into its proper column. A "group" variable is created to help separate the "ANY MEDICATION" row from the other true medications. Percentages are calculated, and the columns (col1–col3) are formatted as "XXX (XXX%)." Finally, the "lastrec" variable is created to help make a continuation flag in the PROC REPORT output.

❺ PROC REPORT is used in this example because there are no special pagination requirements and the pages can break wherever they need to. Note that the &n1, &n2, and &n3 macro variables are used to place the total patient counts into the column headings. The BREAK statement on the group variable ensures division between the "ANY MEDICATION" row and the rest of the table. In order to get a continuation note at the bottom of the table, a COMPUTE block is used with the "lastrec" variable.

Program 5.5 produces the following output.

```
                          Table 5.5
               Summary of Concomitant Medication


         ----------------------------------------------------------------
                           Active          Placebo         Total
         Preferred Medication Term   N= 45          N= 25           N= 70
         ----------------------------------------------------------------
         ANY MEDICATION            11 ( 24%)       10 ( 40%)       21 ( 30%)

         ACETYLSALICYLIC ACID       1 (  2%)        0 (  0%)        1 (  1%)
         DIPHENHYDRAMINE            0 (  0%)        1 (  4%)        1 (  1%)
         FOLIC ACID                 1 (  2%)        0 (  0%)        1 (  1%)
         GABAPENTIN                 1 (  2%)        0 (  0%)        1 (  1%)
         HEPARIN                    1 (  2%)        0 (  0%)        1 (  1%)
         HYDROCORTISONE             2 (  4%)        0 (  0%)        2 (  3%)
         IBUPROFEN                  1 (  2%)        2 (  8%)        3 (  4%)
         IRON                       1 (  2%)        0 (  0%)        1 (  1%)
         --------------------------------------------------------(Continued)
              Created by C:\t_conmeds.sas on 18JAN2005.

<PAGE BREAK HERE...>
                          Table 5.5
               Summary of Concomitant Medication


         ----------------------------------------------------------------
                           Active          Placebo         Total
         Preferred Medication Term   N= 45          N= 25           N= 70
         ----------------------------------------------------------------
         LORAZEPAM                  0 (  0%)        1 (  4%)        1 (  1%)
         MAGNESIUM SULFATE          0 (  0%)        1 (  4%)        1 (  1%)
         MULTIVITAMIN               0 (  0%)        1 (  4%)        1 (  1%)
         NICOTINE                   1 (  2%)        0 (  0%)        1 (  1%)
         POTASSIUM                  1 (  2%)        1 (  4%)        2 (  3%)
         RINGER-LACTATE SOLUTION    0 (  0%)        1 (  4%)        1 (  1%)
         SALMETEROL                 1 (  2%)        0 (  0%)        1 (  1%)
         SODIUM BICARBONATE         0 (  0%)        1 (  4%)        1 (  1%)
         VICODIN                    0 (  0%)        1 (  4%)        1 (  1%)
         ----------------------------------------------------------------
              Created by C:\t_conmeds.sas on 18JAN2005.
```

Creating a Laboratory Shift Table

A laboratory shift table is a tabular display that can show you how a population's laboratory data change, or "shift," over time. Often you want to see what happens to the patients' lab values after therapeutic intervention. Did certain lab parameters drop below or above normal range? Are there laboratory tests that have become of clinical concern? A shift table can provide this information at a glance. Although the example that follows is focused on laboratory data, a shift table can be used to show the movement of any categorical data over time.

The following is a table specification for a laboratory normal range shift table. In order to create this table, you need to have a laboratory data set where the lab values have been flagged as "normal," "low," or "high." The highlighted items in the table shell are parameters that change for the laboratory data in the study.

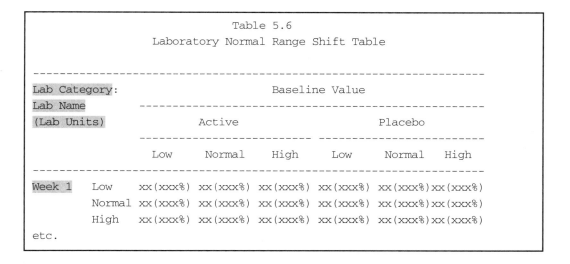

```
                                Table 5.6
                    Laboratory Normal Range Shift Table


       ----------------------------------------------------------------
       Lab Category:                      Baseline Value
       Lab Name         -----------------------------------------------
       (Lab Units)             Active                    Placebo
                        ------------------------  ----------------------
                          Low    Normal    High    Low    Normal   High
       ----------------------------------------------------------------
       Week 1     Low    xx(xxx%) xx(xxx%) xx(xxx%) xx(xxx%) xx(xxx%)xx(xxx%)
                  Normal xx(xxx%) xx(xxx%) xx(xxx%) xx(xxx%) xx(xxx%)xx(xxx%)
                  High   xx(xxx%) xx(xxx%) xx(xxx%) xx(xxx%) xx(xxx%)xx(xxx%)
       etc.
```

The following example relies on a little DATA step programming, a PROC FREQ, and a complicated DATA _NULL_ step for table presentation. Here are the lab shift table annotated SAS program, notes for the program, and output.

Program 5.6 Laboratory Shift Table

```
****  INPUT SAMPLE TREATMENT DATA;
data treat;
label subjid      = "Subject Number"
      trtcd        = "Treatment";
input subjid trtcd @@;
datalines;
101 1   102 0   103 0   104 1   105 0   106 0   107 1   108 1   109 1   110 1
;

****  INPUT SAMPLE LABORATORY DATA;
data lb;
label subjid      = "Subject Number"
      week         = "Week Number"
      lbcat        = "Category for Lab Test"
      lbtest       = "Laboratory Test"
      lbstresu     = "Standard Units"
      lbstresn     = "Numeric Result/Finding in Std Units"
      lbstnrlo     = "Normal Range Lower Limit in Std Units"
      lbstnrhi     = "Normal Range Upper Limit in Std Units"
      lbnrind      = "Reference Range Indicator";

input subjid 1-3 week 6 lbcat $ 9-18 lbtest $ 20-29
      lbstresu $ 32-35 lbstresn 38-41 lbstnrlo 45-48
      lbstnrhi 52-55 lbnrind $ 59;
datalines;
101  0  HEMATOLOGY HEMATOCRIT   %    31    35    49    L
101  1  HEMATOLOGY HEMATOCRIT   %    39    35    49    N
101  2  HEMATOLOGY HEMATOCRIT   %    44    35    49    N
101  0  HEMATOLOGY HEMOGLOBIN  g/dL  11.5  11.7  15.9  L
101  1  HEMATOLOGY HEMOGLOBIN  g/dL  13.2  11.7  15.9  N
101  2  HEMATOLOGY HEMOGLOBIN  g/dL  14.3  11.7  15.9  N
102  0  HEMATOLOGY HEMATOCRIT   %    39    39    52    N
102  1  HEMATOLOGY HEMATOCRIT   %    39    39    52    N
102  2  HEMATOLOGY HEMATOCRIT   %    44    39    52    N
102  0  HEMATOLOGY HEMOGLOBIN  g/dL  11.5  12.7  17.2  L
102  1  HEMATOLOGY HEMOGLOBIN  g/dL  13.2  12.7  17.2  N
102  2  HEMATOLOGY HEMOGLOBIN  g/dL  18.3  12.7  17.2  H
103  0  HEMATOLOGY HEMATOCRIT   %    50    35    49    H
103  1  HEMATOLOGY HEMATOCRIT   %    39    35    49    N
103  2  HEMATOLOGY HEMATOCRIT   %    55    35    49    H
103  0  HEMATOLOGY HEMOGLOBIN  g/dL  12.5  11.7  15.9  N
103  1  HEMATOLOGY HEMOGLOBIN  g/dL  12.2  11.7  15.9  N
103  2  HEMATOLOGY HEMOGLOBIN  g/dL  14.3  11.7  15.9  N
104  0  HEMATOLOGY HEMATOCRIT   %    55    39    52    H
104  1  HEMATOLOGY HEMATOCRIT   %    45    39    52    N
```

```
104    2    HEMATOLOGY HEMATOCRIT    %       44      39      52      N
104    0    HEMATOLOGY HEMOGLOBIN    g/dL    13.0    12.7    17.2    N
104    1    HEMATOLOGY HEMOGLOBIN    g/dL    13.3    12.7    17.2    N
104    2    HEMATOLOGY HEMOGLOBIN    g/dL    12.8    12.7    17.2    N
105    0    HEMATOLOGY HEMATOCRIT    %       36      35      49      N
105    1    HEMATOLOGY HEMATOCRIT    %       39      35      49      N
105    2    HEMATOLOGY HEMATOCRIT    %       39      35      49      N
105    0    HEMATOLOGY HEMOGLOBIN    g/dL    13.1    11.7    15.9    N
105    1    HEMATOLOGY HEMOGLOBIN    g/dL    14.0    11.7    15.9    N
105    2    HEMATOLOGY HEMOGLOBIN    g/dL    14.0    11.7    15.9    N
106    0    HEMATOLOGY HEMATOCRIT    %       53      39      52      H
106    1    HEMATOLOGY HEMATOCRIT    %       50      39      52      N
106    2    HEMATOLOGY HEMATOCRIT    %       53      39      52      H
106    0    HEMATOLOGY HEMOGLOBIN    g/dL    17.0    12.7    17.2    N
106    1    HEMATOLOGY HEMOGLOBIN    g/dL    12.3    12.7    17.2    L
106    2    HEMATOLOGY HEMOGLOBIN    g/dL    12.9    12.7    17.2    N
107    0    HEMATOLOGY HEMATOCRIT    %       55      39      52      H
107    1    HEMATOLOGY HEMATOCRIT    %       56      39      52      H
107    2    HEMATOLOGY HEMATOCRIT    %       57      39      52      H
107    0    HEMATOLOGY HEMOGLOBIN    g/dL    18.0    12.7    17.2    N
107    1    HEMATOLOGY HEMOGLOBIN    g/dL    18.3    12.7    17.2    H
107    2    HEMATOLOGY HEMOGLOBIN    g/dL    19.2    12.7    17.2    H
108    0    HEMATOLOGY HEMATOCRIT    %       40      39      52      N
108    1    HEMATOLOGY HEMATOCRIT    %       53      39      52      H
108    2    HEMATOLOGY HEMATOCRIT    %       54      39      52      H
108    0    HEMATOLOGY HEMOGLOBIN    g/dL    15.0    12.7    17.2    N
108    1    HEMATOLOGY HEMOGLOBIN    g/dL    18.0    12.7    17.2    H
108    2    HEMATOLOGY HEMOGLOBIN    g/dL    19.1    12.7    17.2    H
109    0    HEMATOLOGY HEMATOCRIT    %       39      39      52      N
109    1    HEMATOLOGY HEMATOCRIT    %       53      39      52      H
109    2    HEMATOLOGY HEMATOCRIT    %       57      39      52      H
109    0    HEMATOLOGY HEMOGLOBIN    g/dL    13.0    12.7    17.2    N
109    1    HEMATOLOGY HEMOGLOBIN    g/dL    17.3    12.7    17.2    H
109    2    HEMATOLOGY HEMOGLOBIN    g/dL    17.3    12.7    17.2    H
110    0    HEMATOLOGY HEMATOCRIT    %       50      39      52      N
110    1    HEMATOLOGY HEMATOCRIT    %       51      39      52      N
110    2    HEMATOLOGY HEMATOCRIT    %       57      39      52      H
110    0    HEMATOLOGY HEMOGLOBIN    g/dL    13.0    12.7    17.2    N
110    1    HEMATOLOGY HEMOGLOBIN    g/dL    18.0    12.7    17.2    H
110    2    HEMATOLOGY HEMOGLOBIN    g/dL    19.0    12.7    17.2    H
;
run;


proc sort
    data = lb;
        by subjid lbcat lbtest lbstresu week;
run;
```

```
proc sort
   data = treat;
      by subjid;
run;

**** MERGE TREATMENT INFORMATION WITH LAB DATA.;
data lb;
   merge treat(in = intreat) lb(in = inlb);
      by subjid;

      **** CHECK TO MAKE SURE TREATMENT DATA EXISTS.;           ❶
      if inlb and not intreat then
        put "WARN" "ING: Missing treatment assignment for subject "
              subjid=;

      if intreat and inlb;
run;

**** CARRY FORWARD BASELINE LABORATORY ABNORMAL FLAG.;
data lb;
   set lb;
      by subjid lbcat lbtest lbstresu week;

      retain baseflag " ";

      **** INITIALIZE BASELINE FLAG TO MISSING.;
      if first.lbtest then
         baseflag = " ";

      **** AT WEEK 0 ASSIGN BASELINE FLAG.;
      if week = 0 then
         baseflag = lbnrind;
run;

proc sort
   data = lb;
      by lbcat lbtest lbstresu week trtcd;
run;

**** GET COUNTS AND PERCENTAGES FOR SHIFT TABLE.
**** WE DO NOT WANT COUNTS FOR WEEK 0 BY WEEK 0 SO WEEK 0 IS
**** SUPPRESSED.;
ods listing close;
ods output CrossTabFreqs = freqs;
proc freq
   data = lb(where = (week ne 0));
      by lbcat lbtest lbstresu week trtcd;
```

```
        tables baseflag*lbnrind;
run;
ods output close;
ods listing;


**** WRITE LAB SHIFT SUMMARY TO FILE USING DATA _NULL_;           ❷
options nodate nonumber;
title1 "Table 5.6";
title2 "Laboratory Shift Table";
title3 " ";
data _null_;
   set freqs end = eof;
      by lbcat lbtest lbstresu week trtcd;

      **** SUPPRESS TOTALS.;
      where baseflag ne '' and lbnrind ne '';

      **** SET OUTPUT FILE OPTIONS.;
      file print titles linesleft = ll pagesize = 50 linesize = 80;

      **** WHEN NEWPAGE = 1, A PAGE BREAK IS INSERTED.;
      retain newpage 0;

      **** WRITE THE HEADER OF THE TABLE TO THE PAGE.;
      if _n_ = 1 or newpage = 1 then
         put @1 "----------------------------------"
             "----------------------------------" /
            @1 lbcat ":" @39 "Baseline Value" /
            @1 lbtest
            @17 "--------------------------------"
               "--------------------" /
            @1 "(" lbstresu ")" @25 "Placebo" @55 "Active" /
            @17 "------------------------   "
               "------------------------" /
            @20 "Low     Normal     High      Low      Normal"
               "    High" /
            @1 "----------------------------------"
               "----------------------------------" /;

      **** RESET NEWPAGE TO ZERO.;
      newpage = 0;

      **** DEFINE ARRAY VALUES WHICH REPRESENTS THE 3 ROWS AND
      **** 6 COLUMNS FOR ANY GIVEN WEEK.;
      array values {3,6} $10 _temporary_;
```

```
**** INITIALIZE ARRAY TO "0(  0%)".;
if first.week then
   do i = 1 to 3;
      do j = 1 to 6;
         values(i,j) = "0(  0%)";
      end;
   end;

**** LOAD FREQUENCY/PRECENTS FROM FREQS DATA SET TO
**** THE PROPER PLACE IN THE VALUES ARRAY.;
values( sum((lbnrind = "L") * 1,(lbnrind = "N") * 2,
            (lbnrind = "H") * 3) ,
         sum((baseflag = "L") * 1,(baseflag = "N") * 2,
            (baseflag = "H") * 3) + (trtcd * 3)) =
   put(frequency,2.) || "(" || put(percent,3.) || "%)";

**** ONCE ALL DATA HAVE BEEN LOADED INTO THE ARRAY FOR THE
**** WEEK, PUT THE DATA ON THE PAGE.;
if last.week then
   do;
      put @1 "Week " week
         @10 "Low"      @18 values(1,1) @27 values(1,2)
                        @36 values(1,3) @46 values(1,4)
                        @55 values(1,5) @64 values(1,6) /
         @10 "Normal" @18 values(2,1) @27 values(2,2)
                        @36 values(2,3) @46 values(2,4)
                        @55 values(2,5) @64 values(2,6) /
         @10 "High"     @18 values(3,1) @27 values(3,2)
                        @36 values(3,3) @46 values(3,4)
                        @55 values(3,5) @64 values(3,6) /;

      **** IF IT IS THE END OF THE FILE, PUT A DOUBLE LINE.;
      if eof then
         put @1 "-----------------------------------"
            "-----------------------------------" /
            "-----------------------------------"
            "-----------------------------------" //
            "Created by %sysfunc(getoption(sysin))"
            "on &sysdate9..";
      **** IF ONLY THE LAST WEEK IN A TEST, THEN PAGE BREAK.;
      else if last.lbtest then
         do;
            put @1 "-----------------------------------"
               "-----------------------------------" /
               @60 "(CONTINUED)" /
                  "Created by %sysfunc(getoption(sysin)) "
```

```
                              "on &sysdate9.."
                         _page_;
                     newpage = 1;
                 end;
             end;
    run;
```

Notes for Program 5.6:

❶ Note the check here. You should make a habit of this kind of defensive data check.

❷ A DATA _NULL_ step is used as the reporting tool of choice here primarily because the output format required is too complex for anything else. Because the "CrossTabFreqs" data set from PROC FREQ does not have all of the necessary items on the same observation, a multidimensional array called "values" is used to load and retain all data elements for the given week. The first dimension of the array represents rows and the second dimension represents columns.

Program 5.6 produces the following output.

```
                                  Table 5.6
                             Laboratory Shift Table


---------------------------------------------------------------------------
HEMATOLOGY :                            Baseline Value
HEMATOCRIT        ---------------------------------------------------------
(% )                     Placebo                        Active
                  ---------------------------   ---------------------------
                  Low      Normal    High       Low     Normal     High
---------------------------------------------------------------------------


Week 1   Low      0( 0%)   0( 0%)   0( 0%)    0( 0%)   0( 0%)   0( 0%)
         Normal   0( 0%)   2( 50%)  2( 50%)   1( 17%)  1( 17%)  1( 17%)
         High     0( 0%)   0( 0%)   0( 0%)    0( 0%)   2( 33%)  1( 17%)

Week 2   Low      0( 0%)   0( 0%)   0( 0%)    0( 0%)   0( 0%)   0( 0%)
         Normal   0( 0%)   2( 50%)  0( 0%)    1( 17%)  0( 0%)   1( 17%)
         High     0( 0%)   0( 0%)   2( 50%)   0( 0%)   3( 50%)  1( 17%)


---------------------------------------------------------------------------
                                                            (CONTINUED)
```

(continued)

(continued)

```
Created by C:\t_labshift.sas on 18JAN2005.


<PAGE BREAK HERE...>
                              Table 5.6
                        Laboratory Shift Table


-------------------------------------------------------------------
HEMATOLOGY :                         Baseline Value
HEMOGLOBIN        -------------------------------------------------
(g/dL )                  Placebo                     Active
                  -------------------------   -------------------------
                   Low    Normal    High       Low    Normal    High
-------------------------------------------------------------------

Week 1   Low     0( 0%)  1( 25%)  0( 0%)   0( 0%)  0( 0%)  0( 0%)
         Normal  1( 25%) 2( 50%)  0( 0%)   1( 17%) 1( 17%) 0( 0%)
         High    0( 0%)  0( 0%)   0( 0%)   0( 0%)  4( 67%) 0( 0%)

Week 2   Low     0( 0%)  0( 0%)   0( 0%)   0( 0%)  0( 0%)  0( 0%)
         Normal  0( 0%)  3( 75%)  0( 0%)   1( 17%) 1( 17%) 0( 0%)
         High    1( 25%) 0( 0%)   0( 0%)   0( 0%)  4( 67%) 0( 0%)


-------------------------------------------------------------------
-------------------------------------------------------------------

Created by C:\t_labshift.sas on 18JAN2005.
```

Creating Kaplan-Meier Survival Estimates Tables

Kaplan-Meier survival estimates created during time-to-event analyses are commonly used in clinical trials. With time-to-event analysis, you are comparing time-to-event distributions between therapies. Part of this analysis requires you to create Kaplan-Meier estimate tables that show the probability of a given event over time for each treatment group. The following is a table specification for Kaplan-Meier survival estimates for death. In this table, we are comparing the cumulative probabilities of dying for three different treatment regimens over time.

```
                                Table 5.7
                 Kaplan-Meier Survival Estimates for Death Over Time

---------------------------------------------------------------------------------------

                 Placebo                    Old Drug                    New Drug
          ----------------------     ----------------------     ----------------------
          Remain Surv    95% CIs     Remain Surv    95% CIs     Remain Surv    95% CIs
     Cum.   at  -ival -----------  Cum.  at  -ival -----------  Cum.  at  -ival -----------
Visit Deaths Risk Prob. Lower Upper Deaths Risk Prob. Lower Upper Deaths Risk Prob. Lower Upper
---------------------------------------------------------------------------------------

Baseline  ###   ###  #.### #.### #.###  ###   ###  #.### #.### #.###  ###   ###  #.### #.### #.###

3 Months  ###   ###  #.### #.### #.###  ###   ###  #.### #.### #.###  ###   ###  #.### #.### #.###

6 Months  ###   ###  #.### #.### #.###  ###   ###  #.### #.### #.###  ###   ###  #.### #.### #.###

1 Year    ###   ###  #.### #.### #.###  ###   ###  #.### #.### #.###  ###   ###  #.### #.### #.###

2 Years   ###   ###  #.### #.### #.###  ###   ###  #.### #.### #.###  ###   ###  #.### #.### #.###

3 Years   ###   ###  #.### #.### #.###  ###   ###  #.### #.### #.###  ###   ###  #.### #.### #.###

4 Years   ###   ###  #.### #.### #.###  ###   ###  #.### #.### #.###  ###   ###  #.### #.### #.###
```

The columns are defined as follows:

- "Visit" represents the applicable visit window. Baseline is day 0, a month is represented by 30.4 days, and a year is represented by 365.25 days.
- "Cum. Deaths" represents the cumulative number of deaths in the given visit window.
- "Remain at Risk" is the number of patients who remain at risk for death at the end of the visit window.
- "Survival Prob." is the last survival probability within the visit window.
- The "95% CIs" are 95% lower and upper confidence intervals on the survival probability.

Here is the SAS program that creates this survival estimate table, followed by notes for the program and then the program output.

Program 5.7 Kaplan-Meier Survival Estimates Table

```
**** INPUT SAMPLE TREATMENT AND TIME TO DEATH DATA.;
data death;
label trt        = "Treatment"
      daystodeath = "Days to Death"
      deathcensor = "Death Censor";
input trt $ daystodeath deathcensor @@;
datalines;
A  52    1    A  825   0    C  693   0    C  981   0
B  279   1    B  826   0    B  531   0    B  15    0
C  1057  0    C  793   0    B  1048  0    A  925   0
C  470   0    A  251   1    C  830   0    B  668   1
B  350   0    B  746   0    A  122   1    B  825   0
A  163   1    C  735   0    B  699   0    B  771   1
C  889   0    C  932   0    C  773   1    C  767   0
A  155   0    A  708   0    A  547   0    A  462   1
B  114   1    B  704   0    C  1044  0    A  702   1
A  816   0    A  100   1    C  953   0    C  632   0
C  959   0    C  675   0    C  960   1    A  51    0
B  33    1    B  645   0    A  56    1    A  980   1
C  150   0    A  638   0    B  905   0    B  341   1
B  686   0    B  638   0    A  872   1    C  1347  0
A  659   0    A  133   1    C  360   0    A  907   1
C  70    0    A  592   0    B  112   0    B  882   1
A  1007  0    C  594   0    C  7     0    B  361   0
B  964   0    C  582   0    B  1024  1    A  540   1
C  962   0    B  282   0    C  873   0    C  1294  0
B  961   0    C  521   0    A  268   1    A  657   0
C  1000  0    B  9     1    A  678   0    C  989   1
A  910   0    C  1107  0    C  1071  1    A  971   0
C  89    0    A  1111  0    C  701   0    B  364   1
B  442   1    B  92    1    B  1079  0    A  93    0
B  532   1    A  1062  0    A  903   0    C  792   0
C  136   0    C  154   0    C  845   0    B  52    0
A  839   0    B  1076  0    A  834   1    A  589   0
A  815   0    A  1037  0    B  832   0    C  1120  0
C  803   0    C  16    1    A  630   0    B  546   0
A  28    1    A  1004  0    B  1020  0    A  75    0
C  1299  0    B  79    0    C  170   0    B  945   0
B  1056  0    B  947   0    A  1015  0    A  190   1
B  1026  0    C  128   1    B  940   0    C  1270  0
A  1022  1    A  915   0    A  427   1    A  177   1
C  127   0    B  745   1    C  834   0    B  752   0
A  1209  0    C  154   0    B  723   0    C  1244  0
C  5     0    A  833   0    A  705   0    B  49    0
B  954   0    B  60    1    C  705   0    A  528   0
A  952   0    C  776   0    B  680   0    C  88    0
```

```
C   23     0     B   776    0     A   667    0     C   155    0
B   946    0     A   752    0     C   1076   0     A   380    1
B   945    0     C   722    0     A   630    0     B   61     1
C   931    0     B   2      0     B   583    0     A   282    1
A   103    1     C   1036   0     C   599    0     C   17     0
C   910    0     A   760    0     B   563    0     B   347    1
B   907    0     B   896    0     A   544    0     A   404    1
A   8      1     A   895    0     C   525    0     C   740    0
C   11     0     C   446    1     C   522    0     C   254    0
A   868    0     B   774    0     A   500    0     A   27     0
B   842    0     A   268    1     B   505    0     B   505    1
;
run;


**** PERFORM LIFETEST AND EXPORT SURVIVAL ESTIMATES TO
**** SURVIVALEST DATA SET.;
ods listing close;
ods output ProductLimitEstimates = survivalest;
proc lifetest
   data = death;

   time daystodeath * deathcensor(0);
   strata trt;
run;
ods output close;
ods listing;

**** CALCULATE VISIT WINDOW (MONTHS/YEARS).;
data survivalest;
   set survivalest;

   if daystodeath = 0 then
      visit = 0;      **** Baseline;
   else if 1 <= daystodeath <= 91 then
      visit = 91;     **** 3 Months;
   else if 92 <= daystodeath <= 183 then
      visit = 183;    **** 6 Months;
   else if 184 <= daystodeath <= 365 then
      visit = 365;    **** 1 Year;
   else if 366 <= daystodeath <= 731 then
      visit = 731;    **** 2 Years;
   else if 732 <= daystodeath <= 1096 then
      visit = 1096;   **** 3 Years;
   else if 1097 <= daystodeath <= 1461 then
      visit = 1461;   **** 4 Years;
   else
```

```
            put "ERR" "OR: event data beyond visit mapping "
               daystodeath = ;
   run;

   proc sort
      data = survivalest;
         by trt visit daystodeath;
   run;

   **** CREATE 95% CONFIDENCE INTERVAL AROUND THE ESTIMATE
   **** AND RETAIN PROPER SURVIVAL ESTIMATE FOR TABLE.;
   data survivalest;
      set survivalest;
         by trt visit daystodeath;

         keep trt visit count left survprob lcl ucl;
         retain count survprob lcl ucl;

         **** INITIALIZE VARIABLES TO MISSING FOR EACH TREATMENT.;
         if first.trt then
            do;
               survprob = .;
               count = .;
               lcl = .;
               ucl = .;
            end;

         **** CARRY FORWARD OBSERVATIONS WITH AN ESTIMATE.;
         if survival ne . then
            do;
               count = failed;
               survprob = survival;
               **** SUPPRESS CONFIDENCE INTERVALS AT BASELINE.;
               if visit ne 0 then
                  do;
                     lcl = survival - (stderr*1.96);
                     ucl = survival + (stderr*1.96);
                  end;
            end;

         **** KEEP ONE RECORD PER VISIT WINDOW.;
         if last.visit;
   run;
```

❷

```
proc sort
   data = survivalest;
      by visit;
run;


**** COLLAPSE TABLE BY TREATMENT.  THIS IS DONE BY MERGING THE
**** SURVIVALEST DATA SET AGAINST ITSELF 3 TIMES.;
data table;
  merge survivalest
          (where = (trt = "A")
           rename = (count = count_a left = left_a
                    survprob = survprob_a lcl = lcl_a ucl = ucl_a))
          survivalest
          (where = (trt = "B")
           rename = (count = count_b left = left_b
                    survprob = survprob_b lcl = lcl_b ucl = ucl_b))
          survivalest
          (where = (trt = "C")
           rename = (count = count_c left = left_c
                    survprob = survprob_c lcl = lcl_c ucl = ucl_c));
      by visit;
run;


**** CREATE VISIT FORMAT USED IN TABLE.;
proc format;
   value visit
      0    = "Baseline"
      91   = "3 Months"
    183  = "6 Months"
    365  = "1 Year"
    731  = "2 Years"
    1096 = "3 Years"
    1461 = "4 Years";
run;


**** WRITE SUMMARY STATISTICS TO FILE USING PROC REPORT.;
options  nonumber nodate missing = " "
         formchar = "|----|+|---+=|-/\<>*";
proc report
   data = table
   nowindows
   spacing = 1
   headline
   ls = 101 ps = 30
   split = "|";
```

```
    columns ("--" visit
            ("Placebo|--" count_a left_a survprob_a
                        ("95% CIs|--" lcl_a ucl_a))
            ("Old Drug|--" count_b left_b survprob_b
                        ("95% CIs|--" lcl_b ucl_b))
            ("New Drug|--" count_c left_c survprob_c
                        ("95% CIs|--" lcl_c ucl_c)) );

    define visit       /order order = internal "Visit" left
                        format = visit.;
    define count_a     /display "Cum. Deaths" width = 6
                        format = 3. center;
    define left_a      /display "Remain at Risk" width = 6
                        format = 3. center spacing = 0;
    define survprob_a /display "Surv- ival Prob." center
                        format = pvalue5.3;
    define lcl_a       /display "Lower" format = 5.3;
    define ucl_a       /display "Upper" format = 5.3;
    define count_b     /display "Cum. Deaths" width = 6
                        format = 3. center;
    define left_b      /display "Remain at Risk" width = 6
                        format = 3. center spacing = 0;
    define survprob_b /display "Surv- ival Prob." center
                        format = pvalue5.3;
    define lcl_b       /display "Lower" format = 5.3;
    define ucl_b       /display "Upper" format = 5.3;
    define count_c     /display "Cum. Deaths" width = 6
                        format = 3. center;
    define left_c      /display "Remain at Risk" width = 6
                        format = 3. center spacing = 0;
    define survprob_c /display "Surv- ival Prob." center
                        format = pvalue5.3;
    define lcl_c       /display "Lower" format = 5.3;
    define ucl_c       /display "Upper" format = 5.3;

    break after visit / skip;

title1 "Table 5.7";
title2 "Kaplan-Meier Survival Estimates for Death Over Time";
footnote1 "Created by %sysfunc(getoption(sysin)) on &sysdate9..";
run;
```

Notes for Program 5.7:

❶ Here are the sample death data for the analyses. Note that the treatment information is already present in this single "death" data set. The variable "deathcensor" = 1 if the patient died on that date. "Deathcensor" = 0 means that the patient did not die, and the associated variable "daystodeath" represents the last day that the subject was known to be alive.

❷ Within each visit window, the number of deaths, survival probability, and associated confidence intervals are obtained whenever a death occurs. The values are retained and are output to the data set once per visit at the last record, where the number of subjects remaining at risk is captured in the "left" variable from the "ProductLimitEstimates" data set.

Program 5.7 produces the following output.

```
                                    Table 5.7
                   Kaplan-Meier Survival Estimates for Death Over Time

-----------------------------------------------------------------------------------------

              Placebo                     Old Drug                     New Drug
      --------------------------   --------------------------   --------------------------
              Remain Surv-  95% CIs        Remain Surv-  95% CIs        Remain Surv-  95% CIs
         Cum.   at  ival ----------   Cum.   at  ival ----------   Cum.   at  ival ----------
Visit  Deaths Risk Prob. Lower Upper Deaths Risk Prob. Lower Upper Deaths Risk Prob. Lower Upper
-----------------------------------------------------------------------------------------

Baseline    0    54 1.000             0    52 1.000             0    61 1.000

3 Months    4    47 0.924 0.853 0.996  4    43 0.919 0.842 0.995  1    52 0.983 0.949 1.016

6 Months    9    40 0.823 0.718 0.928  6    40 0.875 0.782 0.969  2    45 0.963 0.914 1.013

1 Year     12    37 0.761 0.643 0.880  7    37 0.853 0.753 0.954  2    44 0.963 0.914 1.013

2 Years    13    24 0.741 0.619 0.862  9    24 0.806 0.692 0.920  3    31 0.942 0.877 1.006

3 Years    16     2 0.522 0.261 0.784 11     0 0.644 0.390 0.897  6     7 0.759 0.546 0.971

4 Years    16     0 0.522 0.261 0.784                            6     0 0.759 0.546 0.971

Created by C:\t_survival.sas on 12MAR2005
```

You can see that the "New Drug" displays better survival probabilities over time than "Old Drug" or "Placebo." You can easily convert this table to a table of Kaplan-Meier failure estimates by replacing "survprob = survival" with "survprob = failure" in Program 5.7.

Creating Listings

Patient listings are usually simple clinical trial patient data displays by data type or domain such as demographics and adverse events. For any given trial report, you will likely have listings of the "raw" data, sometimes called case report form tabulations (CRTs), and you will have listings of the analysis data sets used for analysis. You may be asked to help produce another special type of listing, called a *"patient profile."* Patient profiles are data listings that are separated not by data domain but by patient. Patient profiles are often used by clinical staff to help them write patient narratives for patients who had serious adverse events. We will not discuss how to create patient profile listings here, as their requirements vary widely from trial to trial.

PROC REPORT, PROC PRINT, and DATA _NULL_ programming are the native SAS methods of producing clinical trial listings. Because the power of PROC REPORT is so much greater than that of its predecessor, PROC PRINT, we focus here on the capabilities of PROC REPORT to create clinical trial listings. If we use the data from Table 5.1 presented at the beginning of this chapter, we can create a simple demographics listing that matches the following specification.

```
                            Listing 5.8
                Demographics and Baseline Characteristics

    -----------------------------------------------------------
    Subject
      ID      Treatment    Gender      Race        Age
    -----------------------------------------------------------
      101     Placebo      Male        Other        37
      102     Active       Female      White        65
      103     Active       Male        Black        32
      104     Placebo      Female      White        23
      105     Active       Male        Other        44
      106     Placebo      Female      White        49
      201     Active       Male        Other        35
      202     Placebo      Female      White        50
     (etc.)
    -----------------------------------------------------------
```

Here is the SAS program that creates this listing, followed by notes for the program and then the program output.

Program 5.8 Listing of Demographic Data Using PROC REPORT

```
****  INPUT SAMPLE DEMOGRAPHICS DATA.;
data demog;
label subjid   = "Subject Number"
      trt      = "Treatment"
      gender   = "Gender"
      race     = "Race"
      age      = "Age";
input subjid trt gender race age @@;
datalines;
101 0 1 3 37  301 0 1 1 70  501 0 1 2 33  601 0 1 1 50  701 1 1 1 60
102 1 2 1 65  302 0 1 2 55  502 1 2 1 44  602 0 2 2 30  702 0 1 1 28
103 1 1 2 32  303 1 1 1 65  503 1 1 1 64  603 1 2 1 33  703 1 1 2 44
104 0 2 1 23  304 0 1 1 45  504 0 1 3 56  604 0 1 1 65  704 0 2 1 66
105 1 1 3 44  305 1 1 1 36  505 1 1 2 73  605 1 2 1 57  705 1 1 2 46
106 0 2 1 49  306 0 1 2 46  506 0 1 1 46  606 0 1 2 56  706 1 1 1 75
201 1 1 3 35  401 1 2 1 44  507 1 1 2 44  607 1 1 1 67  707 1 1 1 46
202 0 2 1 50  402 0 2 2 77  508 0 2 1 53  608 0 2 2 46  708 0 2 1 55
203 1 1 2 49  403 1 1 1 45  509 0 1 1 45  609 1 2 1 72  709 0 2 2 57
204 0 2 1 60  404 1 1 1 59  510 0 1 3 65  610 0 1 1 29  710 0 1 1 63
205 1 1 3 39  405 0 2 1 49  511 1 2 2 43  611 1 2 1 65  711 1 1 2 61
206 1 2 1 67  406 1 1 2 33  512 1 1 1 39  612 1 1 2 46  712 0 . 1 49
;

proc sort
   data = demog;
      by subjid;
run;

***** LASTREC VARIABLE IS USED FOR CONTINUING FOOTNOTE.;
data demog;
   set demog end = eof;                                      ❶

   **** FLAG THE LAST OBSERVATION IN THE DATA SET.;
   if eof then
      lastrec = 1;
run;

**** CREATE FORMATS NEEDED FOR LISTING.;
proc format;
   value trt
      1 = "Active"
      0 = "Placebo";
```

```
      value gender
         1 = "Male"
         2 = "Female";
      value race
         1 = "White"
         2 = "Black"
         3 = "Other";
run;

**** USE PROC REPORT TO WRITE LISTING OF DEMOGRAPHICS.;
options formchar="|----|+|---+=|-/\<>*" ls=75 ps=20
         missing = " " nodate nonumber;
proc report
   data = demog
   split = "|"                                              ❷
   spacing = 3
   missing
   nowindows
   headline;

                                                            ❸
   columns ("--" lastrec subjid trt gender race age);

   define lastrec  /display noprint;
   define subjid  /order center width = 7  "Subject|ID" f = 3.;
   define trt     /display left width = 10 "Treatment" f = trt.;
   define gender  /display center width = 10 "Gender"
                   f = gender.;
   define race    /display center width = 10 "Race" f = race.;
   define age     /display center width = 10 "Age" f = 3.;

   **** COMPUTE BLOCK TO PUT CONTINUING TEXT TO PAGE.;
   compute after _page_ / left;                             ❹
   if not lastrec then
      contline = "(Continued)";
   else
      contline = "-----------";

   line @9 "----------------------------------------------"
          contline $11.;
   endcomp;

   title1 "Listing 5.8";
   title2 "Subject Demographics";
   footnote1 "Created by %sysfunc(getoption(sysin)) on"
          " &sysdate9..";
run;
```

Notes for Program 5.8:

❶ The "lastrec" variable is created here to help produce a "continued" footnote for the listing.

❷ Note the very important MISSING option in the PROC REPORT. This ensures that all of your data actually appear in the printed output. Without the MISSING option, PROC REPORT drops any observation with a missing GROUP, ORDER, or ACROSS variable. The NOWINDOWS option is used because this program is designed to run in batch mode and we do not want to run PROC REPORT interactively.

❸ Notice the "headline" PROC REPORT option and the "- -" at the start of the columns statement. This creates a line below and above the column headings.

❹ This is the compute block that places a "continued" footnote at the end of each page unless it has reached the last record in the listing.

Program 5.8 produces the following output.

```
                              Listing 5.8
                          Subject Demographics


        ------------------------------------------------------------
        Subject
          ID      Treatment      Gender        Race          Age
        ------------------------------------------------------------
          101     Placebo        Male          Other          37
          102     Active         Female        White          65
          103     Active         Male          Black          32
          104     Placebo        Female        White          23
          105     Active         Male          Other          44
          106     Placebo        Female        White          49
          201     Active         Male          Other          35
          202     Placebo        Female        White          50
          203     Active         Male          Black          49
          204     Placebo        Female        White          60
        -------------------------------------------------(Continued)


        Created by C:\1_demog.sas on 02MAR2005.

<PAGE BREAK HERE...>
```

(continued)

(continued)

```
                               Listing 5.8
                          Subject Demographics
        -------------------------------------------------------------
        Subject
          ID       Treatment       Gender         Race           Age
        -------------------------------------------------------------
          205      Active          Male           Other          39
          206      Active          Female         White          67
          301      Placebo         Male           White          70
          302      Placebo         Male           Black          55
          303      Active          Male           White          65
          304      Placebo         Male           White          45
          305      Active          Male           White          36
          306      Placebo         Male           Black          46
          401      Active          Female         White          44
          402      Placebo         Female         Black          77
        ------------------------------------------------(Continued)


        Created by C:\l_demog.sas on 02MAR2005.


<PAGE BREAK HERE...>
                               Listing 5.8
                          Subject Demographics
        -------------------------------------------------------------
        Subject
          ID       Treatment       Gender         Race           Age
        -------------------------------------------------------------
          403      Active          Male           White          45
          404      Active          Male           White          59
          405      Placebo         Female         White          49
          406      Active          Male           Black          33
          501      Placebo         Male           Black          33
          502      Active          Female         White          44
          503      Active          Male           White          64
          504      Placebo         Male           Other          56
          505      Active          Male           Black          73
          506      Placebo         Male           White          46
        ------------------------------------------------(Continued)
```

(continued)

(continued)

```
        Created by C:\l_demog.sas on 02MAR2005.

<PAGE BREAK HERE...>


                            Listing 5.8
                        Subject Demographics


        -----------------------------------------------------------
        Subject
          ID       Treatment      Gender        Race          Age
        -----------------------------------------------------------

          507      Active         Male          Black          44
          508      Placebo        Female        White          53
          509      Placebo        Male          White          45
          510      Placebo        Male          Other          65
          511      Active         Female        Black          43
          512      Active         Male          White          39
          601      Placebo        Male          White          50
          602      Placebo        Female        Black          30
          603      Active         Female        White          33
          604      Placebo        Male          White          65
        ------------------------------------------------(Continued)


        Created by C:\l_demog.sas on 02MAR2005.

<PAGE BREAK HERE...>
                            Listing 5.8
                        Subject Demographics


        -----------------------------------------------------------
        Subject
          ID       Treatment      Gender        Race          Age
        -----------------------------------------------------------

          605      Active         Female        White          57
          606      Placebo        Male          Black          56
          607      Active         Male          White          67
          608      Placebo        Female        Black          46
          609      Active         Female        White          72
```

(continued)

(continued)

```
          610      Placebo       Male         White         29
          611      Active        Female       White         65
          612      Active        Male         Black         46
          701      Active        Male         White         60
          702      Placebo       Male         White         28
          ------------------------------------------------(Continued)


     Created by C:\l_demog.sas on 02MAR2005.


<PAGE BREAK HERE...>


                            Listing 5.8
                        Subject Demographics


     --------------------------------------------------------
     Subject
        ID      Treatment      Gender        Race         Age
     --------------------------------------------------------
          703      Active        Male         Black         44
          704      Placebo       Female       White         66
          705      Active        Male         Black         46
          706      Active        Male         White         75
          707      Active        Male         White         46
          708      Placebo       Female       White         55
          709      Placebo       Female       Black         57
          710      Placebo       Male         White         63
          711      Active        Male         Black         61
          712      Placebo                    White         49

     --------------------------------------------------------


     Created by C:\l_demog.sas on 02MAR2005.
```

Note that when sending output to plain ASCII text, PROC REPORT does a fine job of repeating order and group variables at the tops of subsequent pages. However, when PROC REPORT sends output to destinations such as ODS RTF, order and group variables are not repeated at the tops of subsequent pages. When you need broad control over page layout, you may need to revert to using a DATA _NULL_ step with PUT statements to create your listing, as shown in Programs 5.3, 5.4, and 5.6.

Output Appearance Options and Issues

In "Guidance for Industry: Providing Regulatory Submissions in Electronic Format – General Considerations," the FDA requests only that you submit documents that print on 8.5×11-inch paper with 1-inch page margins. The FDA also prefers that you use 12-point font and use a smaller font only if absolutely necessary. So, if you are producing output for the FDA, there are only a few restrictions on your output. However, the FDA is not the only customer of statistical reporting, and some customers demand more attractive output. You may get requests for output requiring specific fonts or formatting that you can get only by using certain proportional fonts. In these cases, the traditional ASCII text provided by the SAS ODS listing destination may not be sufficient. In this section we look at SAS output options for tables and listings and discuss solutions to common problems. The asumption here is that you are producing your tables and listings with either PROC REPORT or a DATA _NULL_ step.

Creating ASCII Text Output

ASCII text is the traditional SAS output found in the SAS LST file, which is the ODS LISTING destination. With ASCII text output you are working with a nonproportional monospace font, and the only real control you have over your output presentation is in using the SAS PAGESIZE and LINESIZE options. However, traditional ASCII text with a monospace font is not a bad option. By using ASCII text with a monospace font you get these benefits:

- You can write extravagant DATA _NULL_ reports that place information on the page wherever you want. You are not limited to rectangular output as you would be with PROC REPORT.

- You can mix complicated DATA _NULL_ reports with more efficient PROC REPORT reports when necessary. Because your output is ASCII text with both types of reports, it is easy to combine these two reporting tools and still have a consistent look and feel to your reporting.

- You can focus more on content and avoid spending time making output "pretty." In other words, ASCII text is usually cheaper to produce.

- PROC REPORT uses more of its features with ASCII text than it does when you send output to other destinations. For instance, order and group variables are repeated at the tops of subsequent pages when you are reporting to the listing destination.

If attractive proportional font output is not required, I recommend that you always do your reporting in SAS with simple ASCII text. It is usually cheaper to produce and more flexible with page layout than any other output destination.

Creating Rich Text Format (RTF) Output

RTF stands for Rich Text Format, which is Microsoft's open markup language for defining the content of documents. If you have an RTF file, simply clicking on it within a Microsoft Windows environment causes Microsoft Word or WordPad to open the file. There are several ways to create RTF files, and thus Microsoft Word files, with SAS.

Sending Output to the ODS RTF Destination

Turning on the SAS ODS RTF destination is the easiest way to send your SAS output to an RTF file. With the following SAS code you can create an RTF file:

```
ods rtf file = "demog_table.rtf";
   proc report
      data = demog;

   etc.
   run;
ods rtf close;
```

Note that adding these two shaded lines of code to your SAS procedure gives you an attractive RTF file. If you do not like the default ODS style template that SAS uses for your output, you can use a different style template supplied by SAS by specifying the STYLE= option in the ODS RTF statement. If you still do not like the way your RTF file looks, you can define your own ODS style template with PROC TEMPLATE. Finally, PROC REPORT allows you to define formatting elements within the SAS procedure itself, where you can customize the appearance of almost any element on the page. For more information about ODS, you can refer to the SAS Press books *Output Delivery System: The Basics*, by Lauren Haworth, and *Quick Results with the Output Delivery System*, by Sunil Gupta. If you just want more style templates to choose from, you can investigate the templates created by Bernadette Johnson in *Instant ODS: Style Templates for the SAS Output Delivery System*.

Although ODS RTF provides a convenient and efficient way to create attractive SAS output, it has a couple of drawbacks:

- SAS TITLE and FOOTNOTE statements translate into Microsoft Word Header and Footer sections when you use ODS RTF. This is fine unless you have a number of ODS RTF files that you need to integrate into a single document, in which case the headers and footers may conflict with one another in your integrated document. You can use the ODS RTF BODYTITLE option to get your titles out of the Microsoft Word header and down into the page space, but that only works if your tables and listings are only one page long. Also, if you use the BODYTITLE option with multi-page output, your titles appear on the first page and your footnotes appear on the last page of your output, which is probably not what you want.

- PROC REPORT does not use all of its features when you use the RTF destination. For instance, ORDER and GROUP variables are not repeated at the tops of subsequent pages.

Other Ways to Convert SAS Output to RTF Files

Besides ODS RTF, there are several other ways to create attractive RTF output from SAS. If you do an Internet search on "RTF SAS macro," you will find SAS macros available that convert a rectangular data set of information into attractive proportional-font RTF files.

There are many ways to take traditional SAS monospace font output and place it into Microsoft Word RTF files. The traditional way to get SAS output into Microsoft Word is to open a Word file and import the SAS LST output manually. Then you adjust the Word margins and font sizes to get your Word document to cooperate with the PAGESIZE and LINESIZE settings that you had in SAS. Some industrious individuals have written elaborate Visual Basic macros in Microsoft Word to import large numbers of SAS LST files automatically into Word. There are also several SAS users group papers on how to convert traditional SAS output into RTF output. Another approach to getting traditional ASCII SAS output into an RTF file is simply to wrap RTF commands around your ASCII SAS output. Here are the steps you would take to do this:

1. Create a blank Microsoft Word document with the page margins, page orientation, and monospace nonproportional font you want.

2. Save the Microsoft Word file as an RTF file.

3. Edit the RTF file with an ASCII text editor such as Notepad, and remove the final curly braces ("}}") at the end of the file. Save this modified RTF file.

4. Concatenate the file you created in step 3, any SAS output you want, and two closing curly braces ("}}") into a file with an "rtf" extension. Note that your SAS output will need the native operating system line feeds and page breaks converted to RTF line feeds and page breaks, which are represented as "\line" and "\page," respectively, in RTF. In UNIX, the SAS code to do the page break conversion might look something like the following:

```
**** convert Unix linefeed to RTF pagebreak.;
if index(text,"^L") then
   text = "\page " || substr(text,2);
```

Rows on the page could just be suffixed with "\line" in a PUT statement to get rows to produce the appropriate line feed.

You could write a general-purpose SAS macro to perform the four preceding steps. Another approach would be to write an operating system script that does this processing for you.

Passing RTF Commands to RTF Output

If you are creating RTF files for your SAS output, you should know that you can pass RTF commands to your output and Microsoft Word will interpret them. For instance, if you want to boldface some text, you can simply preface the text with "\b" and suffix the text to be bolded with "\b0". The full dictionary of RTF commands and the complete RTF specification can be found on the Microsoft Developer Network at http://msdn.microsoft.com. Several SAS users group papers provide the most common RTF commands that you might want.

Note that if you use SAS ODS RTF, you may need to specify "PROTECTSPECIALCHARS=OFF" or define an ODS ESCAPECHAR for your RTF commands to be passed effectively to your RTF file. More details about this can be found in the PROC TEMPLATE FAQ for ODS RTF found at http://support.sas.com/rnd/base/topics/templateFAQ/Template_rtf.html#escapechar.

Creating Portable Document Format (PDF) Files

PDF stands for Portable Document Format, which is an open document description language created by Adobe Systems Inc. PDF files have several advantages over RTF and ASCII text files:

1. PDF files can represent graphics as well as text, which ASCII cannot do.

2. PDF files can be read with the free Adobe Reader. You do not have to purchase a word processing package to open PDF files.

3. Large PDF files are stable. This cannot be said of Microsoft Word documents, which become unstable when files get large.

4. PDF files look the same no matter where you open them. Sometimes Microsoft Word documents can change in appearance depending on your printer or whether you are printing with a PostScript or PCL printer driver.

5. PDF files are the standard for electronic submission of documents to the FDA.

There are several ways to create PDF files from SAS output. Converting a single piece of SAS output into a single PDF file is one task, while converting multiple SAS outputs into a single PDF file is a different task altogether.

Converting a Single SAS Output to PDF

Turning on the SAS ODS PDF destination is the easiest way to send your SAS output to a PDF file. With the following code you can create a PDF file:

```
ods pdf file = "demog_table.pdf";
   proc report
      data = demog;

   etc.
   run;
ods pdf close;
```

Note that adding the two shaded lines of code to the SAS procedure gives you an attractive PDF file. If you do not like the default ODS style template that SAS uses for your output, you can use a different style template supplied by SAS by specifying the STYLE= option in the ODS PDF statement. If you still do not like the way your PDF file looks, you can define your own ODS style template with PROC TEMPLATE. Finally, PROC REPORT allows you to define formatting elements within the SAS procedure itself, where you can customize the appearance of almost any element on the page.

Although ODS PDF provides a convenient and efficient way to create attractive SAS output, PROC REPORT does not use all of its features when you use the PDF destination. For instance, ORDER and GROUP variables are not repeated at the tops of subsequent pages when you send PROC REPORT output to the PDF destination.

There are other ways to convert a single piece of SAS output to PDF. You can use a free script found on the Internet, such as "ascii2pdf," or purchase a PDF file converter from a third-party software vendor. If you have Acrobat Distiller with Acrobat Version 5, you can use the PDF distiller macros in Microsoft Office to print a single file from Microsoft Word to PDF.

Converting or Integrating Multiple SAS Outputs to a Single PDF

If you need to convert multiple SAS output files into a single PDF file, I advise that you use Acrobat Distiller for this purpose. Here are the steps to follow if you have Acrobat Version 5:

1. Start by making PostScript files from your various SAS outputs. If you have ASCII files from SAS, you can use something simple, such as "ascii2ps," to convert your ASCII text to PostScript. You could also create PostScript files directly in SAS by using ODS PS, which is a full-featured PostScript destination. If you have ODS RTF files for output, you can use Microsoft Word as a PostScript file converter to "print" those files "to file" using a generic PostScript printer driver. To convert a large

number of RTF files to PostScript files, it is easy enough to write a Visual Basic macro to print the contents of an entire directory to PostScript files.

2. Use the "Watched Folders" facility, or better yet, write an Acrobat Distiller script to integrate all of your PostScript files into a single PDF. Acrobat Distiller provides a sample script for integrating PostScript files.

If you need to repeatedly regenerate your PDF file, as is often the case in clinical trial reporting, it is easier to integrate PostScript files into a single PDF file with an automated scripted process such as the one listed above. Integrating multiple PDF files into a single PDF file is possible, but it is a manual process within Adobe Acrobat.

"Page X of N" Pagination Solutions

A common requirement for clinical trial output is to have a simple page counter on the page that shows what page you are looking at and the total number of pages. I call this the "Page X of N" pagination problem. There are several ways to solve it. If you are using the ODS listing destination for your SAS output, you can post-process that text in SAS to get a "Page X of N" on the page. If you search the SAS-L listserv, the SAS users group conference proceedings, or http://support.sas.com for "pageof," you are certain to find a SAS macro that suits your needs for output pagination.

Another extremely fast solution to this pagination problem is to write a little Perl script to do the "Page X of N" pagination. Assuming you have Perl, the following script will replace the text "PAGEHERE" placed in your SAS output with a TITLE, FOOTNOTE, or PUT statement with "Page X of N." After you save this script to a file called "pagexofn," you can call the script by entering the command "perl pagexofn sasfile.lst," where "sasfile.lst" is the name of your SAS output file. You can also call this script from within your SAS job by spawning out with the SAS X statement like this: X "perl pagexofn sasfile.lst";

```perl
# open the files.  _temp_output_file_ is a temporary file
open(LSTFILE,$ARGV[0]) or die "Can't open lst file: $!\n";
open(OUTFILE,">_temp_output_file_") or die "Can't open out file:
$!\n";

# get total number of pages in text file
$pagetotal = 0;
while ($line = <LSTFILE>)
{
   if ($line =~ /PAGEHERE/)
   {
      $pagetotal++;
   }
}
```

```
open(LSTFILE,$ARGV[0]) or die "Can't open lst file: $!\n";
# pagenumber = PAGEHERE counter.
$pagenumber = 0;

# increment pagenumber and write the temporary _temp_output_file_
while ($line = <LSTFILE>)
{
   if ($line =~ /PAGEHERE/)
   {
      $pagenumber++;
   }
   $line =~ s/PAGEHERE/$pagenumber of $pagetotal/;
   print OUTFILE "$line";
}

# overwrite the source text file with the temporary file
use File::Copy;
move("_temp_output_file_",$ARGV[0]) or die "move failed $!";
```

If you are using ODS RTF for your output destination, you can pass the RTF command string for "Page X of N" directly to your SAS output like this:

```
ods rtf file = "demog_table.rtf";
  proc report
    data = demog;
    ...
    footnote1 "{\cgrid0 PAGE }{\field{\*\fldinst {\cs17  PAGE
}}{\fldrslt {\cs17\lang10241}}}{\cs17  OF }{\field{\*\fldinst {\cs17
NUMPAGES }}{\fldrslt{\cs17\lang1024 1}}}";
  run;
ods rtf close;
```

The shaded text is one big contiguous text string. You can change the boldface text to something different if you prefer.

Footnote Indicating SAS Program and Date

Another common request for SAS output is that it have a footnote showing the name of the program that produced it and when it was produced. In the UNIX environment, you can use the following SAS footnote to get the program and datetime stamp:

```
footnote1
"Created by %sysget(PWD)/%sysfunc(scan(&SYSPROCESSNAME,2)).sas
generated on &sysdate9 &systime" ;
```

This generates a footnote in your SAS output that might look something like this:

```
Created by /trialname/demog.sas generated on 09AUG2004 10:57
```

The Microsoft Windows footnote would look like this:

```
footnote1 "Created by %sysfunc(getoption(sysin)) on &sysdate9
&systime";
```

Please note that the Microsoft Windows footnote populates SYSIN only if the program is run in batch mode.

SAS Macro-Based Reporting Systems

Many pharmaceutical companies and CROs have developed and use a SAS macro-based reporting system to generate clinical trial tables and listings. This is because many of the tasks associated with table and listing production are repetitive. As stated in Chapter 1, you should strive to "program it once," and a SAS macro reporting system allows you to do that. Some SAS macro clinical trial reporting tools can even be found for free on the Internet, and other clinical trial macro reporting systems are available for purchase.

Building your own SAS macro reporting system is not conceptually difficult. You start by looking for common tasks across your clinical trial programming. Then, you design SAS macro code that parameterizes that programming. For instance, in Table 5.3 in this chapter, the SAS code that summarizes gender and race are almost identical. A simple SAS macro that accepted the categorical variable as a SAS macro parameter would have been a good solution there. Here is a small list of common reporting tasks that you can look toward generalizing with the SAS macro language:

1. Categorical data analysis

2. Continuous data analysis

3. Data transposition

4. Output formatting, pagination

I recommend that you follow a more traditional systems development life-cycle model (SDLC) when developing a comprehensive SAS macro-based reporting system. Unlike much of the one-time-only SAS programming that occurs for a clinical trial, you need to ensure that a general-purpose SAS macro system is robust enough to handle any problem it encounters. The systems development life-cycle approach to software development will help you to build strong software applications with SAS.

C h a p t e r 6

Creating Clinical Trial Graphs

Common clinical trial graphics are the focus of this chapter. First, we discuss the types of graphs that are most often encountered in clinical trial analysis and reporting. Then we examine the various tools that SAS provides to help produce these graphs. Sample graph programs are provided to show how many of these graphs can be produced.

Common Clinical Trial Graphs

There are several types of graphs that are common to clinical trial analysis and reporting. What follows are some brief descriptions of these graphs.

Scatter Plot

A *scatter plot* provides a quick visual summary of where numerous data points exist in two-dimensional space. Scatter plots are commonly used to investigate positive or negative *correlation* between two variables plotted on the X and Y axes.

Figure 6.1 Sample Scatter Plots

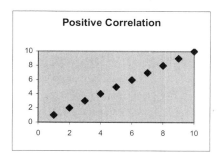
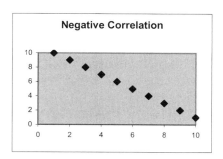

In this case, you see in the graph on the left that with each increase in X there is an equivalent increase in Y. Those are positively correlated variables. In the graph on the right, each increase of X is associated with an equivalent decrease in Y. Those are negatively correlated variables.

Scatter plots are used widely in clinical trial research, as they are intuitive to read and have many applications. We often look for a drug's effect on some parameter (Y axis) over time (X axis). Also, *change-from-baseline scatter plots* are useful when you plot the baseline on the X axis and the follow-up value on the Y axis. You will see an example of this graph later in this chapter.

Line Plot

Another commonly used graph in clinical trials is the *line plot*. A line plot lets you display the count or measure of something on the Y axis across some other dimension on the X axis. The following is a sample line plot where each line represents a treatment regimen and the treatment response (Y axis) is measured across time (X axis).

Figure 6.2 Sample Line Plot

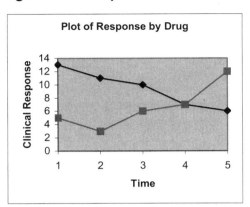

Bar Chart

Bar charts can represent data similar to the data represented by a line plot, but in a visually different way. Sometimes bar charts are requested in clinical trial reporting instead of a line plot. The following example shows the data from the previous line plot represented as a bar chart.

Figure 6.3 Sample Bar Chart

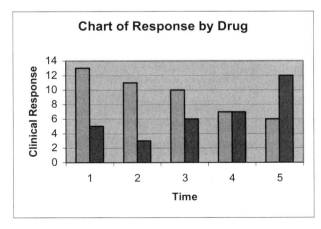

Bar charts can be represented either horizontally or vertically depending on user preference, but you can represent the same data and information in a bar chart as you can in a line plot.

Box Plot

Box plots, also known as box and whisker plots, are commonly used to display univariate statistics for a given variable across another variable. The statistics typically displayed in a box plot are the minimum, first quartile, median, third quartile, and maximum values. Mean values are often included in box plots as well. The following is a sample box plot of a clinical response measure showing how three different drug therapies compare to one another.

Figure 6.4 Sample Box Plot

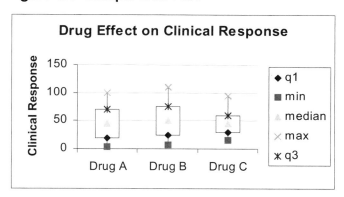

Odds Ratio Plot

As a part of *logistic regression* analysis, *odds ratio plots* are an excellent way to see how much more likely a condition is to exist based on the presence of another condition. Just by glancing at an odds ratio plot, you can see whether an independent variable is significant to the dependent variable. For instance, if the odds ratio confidence interval does not cross the value of 1, then the independent variable odds ratio is significant. Examine the following graph.

Figure 6.5 Sample Odds Ratio Plot

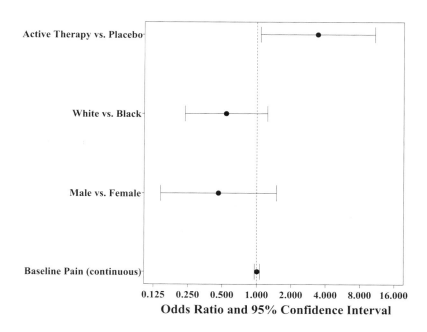

In this example, we see that "Active Therapy vs. Placebo," or drug therapy, has a significant odds ratio because the 95% confidence interval line does not cross 1. It appears that patients on active therapy are almost four times as likely to experience clinical success as those who are not on active therapy, while controlling for the variables "White vs. Black," "Male vs. Female," and "Baseline Pain (continuous)."

Kaplan-Meier Survival Estimates Plot

Comparison of *Kaplan-Meier survival estimates* is often called for in clinical trial analysis. With *survival analysis*, you are trying to determine which treatment group displays a better "*time-to-event*" distribution than another. Part of this analysis is the production of Kaplan-Meier estimates plots that show the probability of a given event over time for each treatment group. In the following example you see that "New Drug" displays better survival estimates over time than either "Old Drug" or "Placebo."

Figure 6.6 Sample Kaplan-Meier Survival Estimates Plot

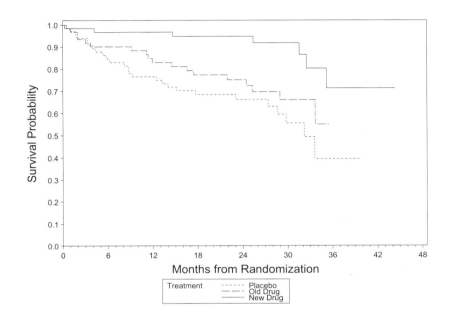

SAS Tools for Creating Clinical Trial Graphs

SAS provides many tools for creating graphical displays of clinical trial data. The majority of these tools can be found in SAS/GRAPH software, and those are what we explore here.

Common Clinical Trial SAS/GRAPH Procedures

There are a number of SAS procedures that help to build the graphics most common to clinical trial reporting. The following table describes which SAS procedures make which graphs.

Type of Plot	SAS Procedure
Scatter plot	PROC GPLOT or PROC REG
Line plot	PROC GPLOT
Bar chart	PROC GCHART
Box plot	PROC BOXPLOT*, PROC GPLOT, or PROC UNIVARIATE
Kaplan-Meier plot	PROC LIFETEST* or PROC GPLOT

* Found in SAS/STAT documentation.

Note that for box and survival plots, PROC GPLOT is listed as an alternative. Although PROC BOXPLOT and PROC LIFETEST produce excellent graphs by themselves, sometimes it is necessary to make modifications to the output in a way that these procedures cannot handle directly. When modifications are needed, PROC GPLOT is an excellent choice. Also note that PROC REG and PROC UNIVARIATE are listed as options for scatter plots and box plots, respectively, as they can be useful in producing lower-resolution graphics for statistical appendices.

Using the Annotate Facility for Graph Augmentation

SAS/GRAPH software is full of SAS procedures that produce a large variety of graphs. If a graph produced by one of the SAS/GRAPH procedures is not exactly what is needed, it can be augmented with the SAS/GRAPH Annotate facility. The Annotate facility provides a way to overlay graphics transparently on a SAS graph. This is done by placing graphics commands inside a special SAS data set called an Annotate data set. The Annotate data set is then fed to a SAS/GRAPH procedure and gets processed along with the SAS/GRAPH code.

Imagination is the only limit to how a SAS graph can be augmented with the Annotate facility. It is an extremely powerful way to customize SAS graphics, and you will see examples of that in this chapter. The chapter does not discuss the stand-alone DATA Step Graphics Interface, which allows you to create SAS graphics using only DATA step commands. Usually a SAS procedure will get you far enough along in creating a graph that a simple set of Annotate facility commands is sufficient to complete the graph.

Sample Graphs

In this section we look at some SAS code examples that create some of the most common clinical trial graphs.

Creating a Scatter Plot

The following example is a scatter plot of hematocrit values by treatment regimen using PROC GPLOT. Included in the plot is a line of equivalence drawn diagonally through the plot using the Annotate facility. By using a scatter plot such as this, you can see how treatment might affect hematocrit values. For instance, if for a given therapy you see that most values fall below the line of equivalence, then you might guess that the therapy contributes to an increase in hematocrit values.

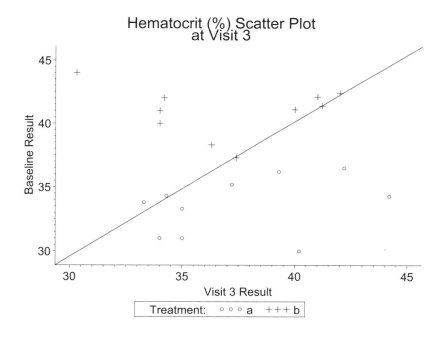

Here is the SAS program that creates the preceding scatter plot. Notes follow the program.

Program 6.1 Laboratory Data Scatter Plot Using PROC GPLOT

```
**** INPUT SAMPLE HEMATOCRIT LAB DATA.;
data labs;
label baseline = "Baseline Result"
      value     = "Visit 3 Result"
      trt       = "Treatment";
input subject_id lbtest $ value baseline trt $ @@;
datalines;
101 hct 35.0 31.0 a     102 hct 40.2 30.0 a
103 hct 42.0 42.4 b     104 hct 41.2 41.4 b
105 hct 35.0 33.3 a     106 hct 34.3 34.3 a
107 hct 30.3 44.0 b     108 hct 34.2 42.0 b
109 hct 40.0 41.1 b     110 hct 41.0 42.1 b
111 hct 33.3 33.8 a     112 hct 34.0 31.0 a
113 hct 34.0 41.0 b     114 hct 34.0 40.0 b
115 hct 37.2 35.2 a     116 hct 39.3 36.2 a
117 hct 36.3 38.3 b     118 hct 37.4 37.3 b
119 hct 44.2 34.3 a     120 hct 42.2 36.5 a
;
run;

**** DEFINE GRAPHICS OPTIONS:  SET DEVICE DESTINATION TO MS
**** OFFICE CGM FILE, REPLACE ANY EXISTING CGM FILE, RESET ANY
**** SYMBOL DEFINITIONS, AND SET COLORS TO BLACK.;
filename fileref1 "C:\lab_scatter.cgm";                      ❶
goptions device  = CGMOF97L
         gsfname = fileref1
         gsfmode = replace
         reset   = symbol
         colors  = (black);

**** DEFINE PLOT SYMBOLS. TRT = A = CIRCLE, B = PLUS SIGN.;
symbol1 value = CIRCLE;
symbol2 value = PLUS;

**** DEFINE VERTICAL AXIS OPTIONS.;                          ❷
axis1 order = (30 to 45 by 5)
      label = (angle = 90 height = 1.5)
      value = (height = 1.5)
      offset = (1 cm, );

**** DEFINE HORIZONTAL AXIS OPTIONS.;
axis2 order = (30 to 45 by 5)
```

```
        label = (height = 1.5)
        value = (height = 1.5)
        offset = (1 cm, );

**** DEFINE THE LEGEND FOR THE BOTTOM OF THE GRAPH.;
legend1 frame
        value = (height = 1.5)
        label = (height = 1.5 justify = right 'Treatment:')
        position = (bottom center outside);


**** CREATE LINE OF EQUIVALENCE USING AN ANNOTATE DATA SET.;
data annotate;

    format function $8.;
    format xsys ysys $1.;

    **** SET COORDINATE SYSTEM AND POSITIONING.
    **** XSYS/YSYS = 1 SETS THE COORDINATES RELATIVE TO THE
    **** PLOT AREA.;
    xsys = '1';
    ysys = '1';
    line = 1;
    size = 2;
    color = 'BLACK';

    **** MOVE TO LOWER LEFT CORNER.;
    function = 'MOVE';
    x = 0;
    y = 0;
    output;

    **** DRAW LINE FROM LOWER LEFT CORNER TO UPPER RIGHT CORNER.;
    function = 'DRAW';
    x = 100;
    y = 100;
    output;
run;

**** CREATE SCATTERPLOT.  BASELINE IS ON THE Y AXIS, VISIT 3 IS
**** ON THE X AXIS, AND THE VALUES ARE PLOTTED BY TREATMENT.;
proc gplot
    data = labs;
```

```
plot baseline * value = trt / vaxis = axis1
                              haxis = axis2
                              anno = annotate
                              legend = legend1
                              noframe;

title1 height = 2 font = "Helvetica"
       "Hematocrit (%) Scatter Plot";
title2 height = 2 font = "Helvetica"
       "at Visit 3";
run;
quit;
```

Notes for Program 6.1:

❶ The very important GOPTIONS statements are specified at this point. For this graph (and for the others in this book) we send the output to a Microsoft Office Computer Graphics Metafile (CGM). At the end of this chapter is a discussion of all the various output options and the benefits and disadvantages of each.

❷ Notice that the vertical axis has had the label text rotated and that the range has been explicitly specified for each axis. It is often the case that an axis has the range explicitly specified, but this comes at a cost. First, if an axis range is explicitly specified, then BY-statement processing of the subsequent PROC GPLOT can be limited. For instance, in this example we would not be able to run the PROC GPLOT "by lbtest" because the axis range would change for each lab test. Second, PROC GPLOT and other procedures truncate data that fall outside the specified axis range. For example, if we had a hematocrit value of 55 in the labs data set, then we would see this note in the SAS log:

```
"NOTE: 1 observation(s) outside the axis range for the
baseline * value = trt request."
```

Creating a Line Plot

Now let's examine a sample line plot that shows the clinical response of a population of patients to two different therapies over the course of 10 days. Here is the sample line plot.

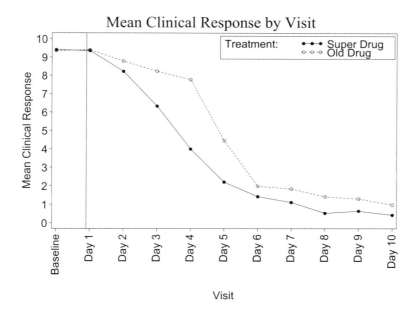

Here is the SAS program that creates this plot.

Program 6.2 Clinical Response Line Plot Using PROC GPLOT

```
**** INPUT SAMPLE MEAN CLINICAL RESPONSE VALUES.;
data response;
label response = "Mean Clinical Response"
      visit    = "Visit"
      trt      = "Treatment";
input trt visit response @@;
datalines;
1  0 9.40    2  0 9.35
1  1 9.35    2  1 9.40
1  2 8.22    2  2 8.78
1  3 6.33    2  3 8.23
1  4 4.00    2  4 7.77
1  5 2.22    2  5 4.46
1  6 1.44    2  6 2.00
1  7 1.13    2  7 1.86
1  8 0.55    2  8 1.44
```

```
1  9 0.67    2  9 1.33
1 10 0.45    2 10 1.01
;
run;

**** CREATE VISIT FORMAT.;
proc format;
   value visit
       0 = "Baseline"
       1 = "Day 1"
       2 = "Day 2"
       3 = "Day 3"
       4 = "Day 4"
       5 = "Day 5"
       6 = "Day 6"
       7 = "Day 7"
       8 = "Day 8"
       9 = "Day 9"
      10 = "Day 10";

   value trt
       1 = "Super Drug"
       2 = "Old Drug";
run;

**** DEFINE GRAPHICS OPTIONS:  SET DEVICE DESTINATION TO MS
**** OFFICE CGM FILE, REPLACE ANY EXISTING CGM FILE, RESET ANY
**** SYMBOL DEFINITIONS, AND SET COLORS TO BLACK.;
filename fileref "C:\plot_signsymp_mean.cgm";
goptions device = CGMOF97L
         gsfname = fileref
         gsfmode = REPLACE
         reset = symbol
         colors = (black);

**** DEFINE PLOT SYMBOLS. DOT = SUPER DRUG WITH SOLID LINE,
**** CIRCLE = OLD DRUG WITH DOTTED LINE.  I = J CONNECTS POINTS.;
symbol1 c = black line = 1 v = dot i = j;
symbol2 c = black line = 2 v = circle i = j;

**** DEFINE HORIZONTAL AXIS OPTIONS.;
axis1 order = (0 to 10 by 1)
      value = (angle = 90 height = 1.5)
      label = (height = 1.5 "Visit")
      minor = none
      offset = (1,1);
```

```
**** DEFINE VERTICAL AXIS OPTIONS.;
axis2 order = (0 to 10 by 1)
      value = (height = 1.5)
      label = (height = 1.5 angle = 90 "Mean Clinical Response")
      minor = none
      offset = (1,1);

**** DEFINE THE LEGEND FOR THE TOP RIGHT OF THE GRAPH.;
legend1 label = (height = 1.5 "Treatment:")
        order = (1 2)
        value = (height = 1.5 justify = left
                 "Super Drug" "Old Drug" )
        down = 2
        position = (top right inside)
        mode = protect
        frame;

**** CREATE LINE PLOT.  VISIT IS ON THE X AXIS, RESPONSE IS ON
**** THE Y AXIS, AND THE VALUES ARE PLOTTED BY TREATMENT.   A
**** REFERENCE LINE IS DRAWN JUST BEFORE DAY 1 TO SET APART
**** POST TREATMENT DATA.;
proc gplot
   data = response;

   plot response * visit = trt /haxis = axis1
                                vaxis = axis2
                                legend = legend1
                                href = (0.9);
   format visit visit.;
   title1 h = 2 font = "TimesRoman"
          "Mean Clinical Response by Visit";
run;
quit;
```

Creating a Bar Chart

The following is an example of a bar chart showing patient pain score ratings by treatment. The percentage of patients with each pain rating is displayed for each drug classification.

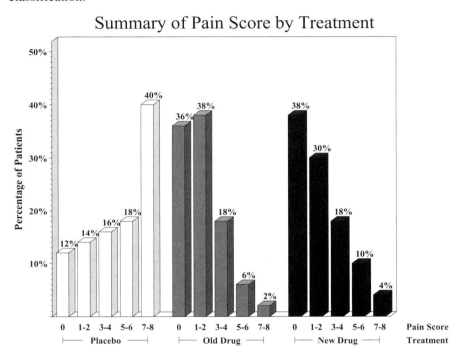

Here is the SAS program that creates this vertical three-dimensional bar chart.

Program 6.3 Clinical Response Bar Chart Using PROC GPLOT

```
**** INPUT SAMPLE PAIN SCALE DATA.;
data pain;
label subject = "Subject"
      pain    = "Pain Score"
      trt     = "Treatment";
input subject pain trt @@;
datalines;
113    1 1         420    1 2         780    0 3
121    1 1         423    0 2         784    0 3
122    1 1         465    4 2         785    1 3
```

124	4	1	481	3	2	786	3	3
164	4	1	482	0	2	787	0	3
177	4	1	483	0	2	789	0	3
178	0	1	484	0	2	790	2	3
179	1	1	485	0	2	791	0	3
184	0	1	486	1	2	793	3	3
185	0	1	487	0	2	794	2	3
186	3	1	489	0	2	795	1	3
187	0	1	490	1	2	796	4	3
188	1	1	491	0	2	797	2	3
189	3	1	493	2	2	798	1	3
190	3	1	494	1	2	799	2	3
191	2	1	495	0	2	800	2	3
192	3	1	496	2	2	822	1	3
193	4	1	498	2	2	841	1	3
194	4	1	499	2	2	842	1	3
195	0	1	500	2	2	847	2	3
196	4	1	521	1	2	863	1	3
197	1	1	522	1	2	881	2	3
198	4	1	541	1	2	966	1	3
199	3	1	542	0	2	967	0	3
100	4	1	561	3	2	968	0	3
121	2	1	562	2	2	981	1	3
122	2	1	581	2	2	982	1	3
123	4	1	561	1	2	985	0	3
124	2	1	564	1	2	986	0	3
141	3	1	566	1	2	987	0	3
142	2	1	567	2	2	989	2	3
143	2	1	568	2	2	990	3	3
147	4	1	569	0	2	991	0	3
161	4	1	581	0	2	992	2	3
162	4	1	582	3	2	993	1	3
163	4	1	584	1	2	994	0	3
164	0	1	585	0	2	995	1	3
165	2	1	586	1	2	996	0	3
166	1	1	587	1	2	997	3	3
167	3	1	591	1	2	998	0	3
181	2	1	592	1	2	999	0	3
221	4	1	594	1	2	706	0	3
281	4	1	595	0	2	707	3	3
282	4	1	596	0	2	708	1	3
361	4	1	597	0	2	709	0	3
362	4	1	601	0	2	710	1	3
364	3	1	602	1	2	711	1	3
365	4	1	603	2	2	712	0	3
366	3	1	604	1	2	713	4	3
367	4	1	605	1	2	714	0	3

;

```
**** MAKE FORMATS FOR CHART.;
proc format;
   value trt
       1 = 'Placebo'
       2 = 'Old Drug'
       3 = 'New Drug';

   value score
       0 = '0'
       1 = '1-2'
       2 = '3-4'
       3 = '5-6'
       4 = '7-8';

   picture newpct (round)
       0 = " "
       0 < - < .5 = "<1%"
       .6 < - high = "000%";
run;

proc sort
   data = pain;
       by trt;
run;

**** GET FREQUENCY COUNTS FOR CHART AND PUT IN FREQOUT DATA SET.;
proc freq
   data = pain
   noprint;
       by trt;

       tables pain /out = freqout;
run;

**** DEFINE GRAPHICS OPTIONS:  SET DEVICE DESTINATION TO MS
**** OFFICE CGM FILE, REPLACE ANY EXISTING CGM FILE, RESET ANY
**** SYMBOL DEFINITIONS, AND SET BACKGROUND TO WHITE AND OTHER
**** COLORS TO BLACK.;
filename fileref "C:\bar_chart.cgm";
goptions device = cgmof971
         gsfname = fileref
         gsfmode = replace
         reset = symbol
         cback = white
         colors = (black)
         chartype = 6;
```

```
**** DEFINE BAR PATTERNS: WHITE = PLACEBO, GRAY = OLD DRUG,
**** BLACK = NEW DRUG.;
pattern1 value = solid color = white;
pattern2 value = solid color = gray;
pattern3 value = solid color = black ;

**** DEFINE HORIZONTAL AXIS OPTIONS.;
axis1 label = (h = 1 'Pain Score')
      value = (h = 1 )
      order = (0 to 4 by 1);

**** DEFINE VERTICAL AXIS OPTIONS.;
axis2 label = (h = 1.2 r = 0 a = 90  'Percentage of Patients' )
      order = (0 to 50 by 10);

**** CREATE BAR CHART.  PERCENTAGE OF PATIENTS IS ON THE Y AXIS,
**** PAIN SCORE BY TREATMENT IS ON THE X AXIS.;
proc gchart
   data = freqout;

   vbar3d pain /group = trt
                sumvar = percent
                maxis = axis1
                raxis = axis2
                midpoints = 0 1 2 3 4
                cframe = white
                coutline = black
                outside = sum
                patternid = group;

   format trt trt.
          pain score.
          percent newpct.;
   title1 font = "TimesRoman" h = 2 color = black
          "Summary of Pain Score by Treatment";
run;
quit;
```

There are other means of creating bar charts in SAS. If you want a quick histogram, you can call PROC UNIVARIATE with the HISTOGRAM option. Also, PROC CAPABILITY with the HISTOGRAM statement may be used to create histograms. PROC UNIVARIATE is explained in detail in the *SAS/STAT User's Guide,* and PROC CAPABILITY is covered in the *SAS/QC User's Guide.*

Creating a Box Plot

The following is an example of a box plot of seizure data by treatment at each of three visits. This first example is created with the traditional PROC GPLOT. This box plot has boxes that span the *interquartile range* and whiskers that extend to the maximum and minimum values. Also, a connecting line joins the median values for each treatment.

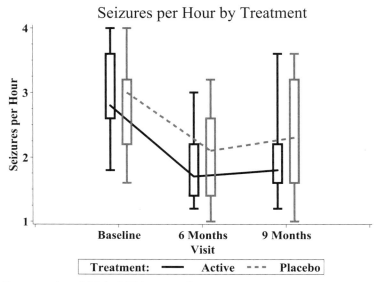

Box extends to 25th and 75th percentile.
Whiskers extend to minimum and maximum values

Here is the SAS program that creates this box plot.

Program 6.4 Creating a Box Plot Using PROC GPLOT

```
**** INPUT SAMPLE PAIN SCALE DATA;
data seizures;
label seizures = "Seizures per Hour"
      visit    = "Visit"
      trt      = "Treatment";
input trt visit seizures @@;
datalines;
1 2   1.5        2    1    3        2    2    1.8
2 1   2.6        2    2    2        2    3    2
1 1   2.8        2    3    2.6      1    1    3
1 2   2.2        1    1    2.4      2    1    3.2
2 1   3.2        1    2    1.4      1    1    2.6
2 2   2.1        1    3    1.8      1    2    1.2
1 1   2.6        2    1    3        1    3    1.8
```

```
2 1    2.2          1    1    3.6          2    1    2.1
2 2    3.2          1    2    2            2    2    1
1 1    2.6          1    3    3.6          2    3    1.8
1 2    2.2          2    1    3.6          1    1    2.6
1 3    2.2          2    2    2.6          2    1    4
2 1    2.8          2    3    2            2    3    3.6
2 2    2.6          1    1    2.8          1    1    3.4
2 3    2.6          1    2    1.8          1    2    3
1 1    2.0          1    3    1.6          2    1    3.4
1 2    2.4          2    1    3.8          2    2    2
2 1    2.1          2    2    3            1    1    2.6
2 2    1.2          2    3    3.4          1    3    1.8
2 3    1            1    1    4            2    1    2.0
1 1    2.9          1    3    3.4          1    1    2.8
1 2    1.6          2    1    2.8          2    1    2.4
1 3    1.2          2    2    1.2          1    1    3.6
2 1    2.8          2    3    1.2          2    1    3.2
2 2    2.6          1    1    1.8          2    2    2.2
2 3    3.2          1    2    2            2    3    3.2
1 1    2.8          1    3    2.2          1    1    4
1 2    1.4          2    1    3            2    1    3.2
1 3    2            2    2    1.4          1    1    2.4
2 1    1.6          2    3    1.4          2    1    4
1 1    2.8          1    1    3.6          2    2    2.2
1 2    1.4          1    2    1.4          1    1    4
1 3    1.2          2    1    2.2
;

proc format;
   value trt
       1 = "Active"
       2 = "Placebo";
run;


**** CREATE PLOTVISIT VARIABLE WHICH IS A SLIGHTLY OFFSET VISIT
**** VALUE TO MAKE TRT DISTINGUISHABLE ON THE X AXIS.  OTHERWISE,
**** TREATMENT 1 AND 2 WOULD HAVE OVERLAPPING BOXES.;
data seizures;
   set seizures;

   if trt = 1 then
      plotvisit = visit - .1;
   else if trt = 2 then
      plotvisit = visit + .1;
run;
```

```
**** DEFINE GRAPHICS OPTIONS:  SET DEVICE DESTINATION TO MS
**** OFFICE CGM FILE, REPLACE ANY EXISTING CGM FILE, RESET ANY
**** SYMBOL DEFINITIONS, AND DEFINE DEFAULT FONT TYPE.;
filename fileref 'C:\whisker.cgm';
goptions
   device = cgmof971
   gsfname = fileref
   gsfmode = replace
   reset = symbol
   colors = (black)
   chartype = 6;

**** SET SYMBOL DEFINITIONS.
**** SET LINE THICKNESS WITH WIDTH AND BOX WIDTH WITH BWIDTH.
**** ACTIVE DRUG IS DEFINED AS A SOLID BLACK LINE IN SYMBOL1
**** AND PLACEBO GETS A DASHED GRAY LINE IN SYMBOL2.
**** VALUE = NONE SUPPRESSES THE ACTUAL DATA POINTS.
**** BOXJT00 MEANS TO CREATE BOX PLOTS WITH BOXES (25TH AND 75TH
**** PERCENTILES - THE INTERQUARTILE RANGE) JOINED (J) AT THE
**** MEDIANS WITH WHISKERS EXTENDING TO THE MINIMUM AND MAXIMUM
**** VALUES (00) AND TOPPED/BOTTOMED (T) WITH A DASH.
**** MODE = INCLUDE OPTION ENSURES THAT VALUES THAT MIGHT FALL
**** OUTSIDE OF THE EXPLICITLY STATED AXIS ORDER WOULD BE
**** INCLUDED IN THE BOX AND WHISKER DEFINITION.;
symbol1 width = 28 bwidth = 3 color = black line = 1 value = none
        interpol = BOXJT00 mode = include;
symbol2 width = 28  bwidth = 3 color = gray line = 2 value = none
        interpol = BOXJT00 mode = include;

**** DEFINE THE LEGEND FOR THE BOTTOM CENTER OF THE PAGE.;
legend1
   frame
   value = (height = 1.5)
   label = (height = 1.5 justify = right 'Treatment:' )
   position = (bottom center outside);

**** DEFINE VERTICAL AXIS OPTIONS.;
axis1 label = (h = 1.5 r = 0 a = 90 "Seizures per Hour")
      value = (h = 1.5 )
      minor = (n = 3);
```

```
**** DEFINE HORIZONTAL AXIS OPTIONS.
**** THE HORIZONTAL AXIS MUST GO FROM 0 TO 4 HERE BECAUSE OF THE
**** OFFSET APPLIED TO VISIT.  NOTICE THAT THE VALUE FOR VISIT
**** OF 0 AND 4 IS SET TO BLANK.;
axis2 label = (h = 1.5  "Visit") .
      value = (h = 1.5 " " "Baseline"  "6 Months" "9 Months" " ")
       order = (0 to 4 by 1)
       minor = none;


**** CREATE BOX PLOT.  VISIT IS ON THE X AXIS, SEIZURES ARE ON
**** THE Y AXIS, AND THE VALUES ARE PLOTTED BY TREATMENT.;
proc gplot
    data = seizures;
    plot seizures * plotvisit = trt  /vaxis = axis1
                                      haxis = axis2
                                      legend = legend1
                                      noframe;
    format trt trt.;

    title1 h = 2 font = "TimesRoman"
          "Seizures per Hour by Treatment";
    footnote1 h = 1.5 j = l font = "TimesRoman"
          "Box extends to 25th and 75th percentile.";
    footnote2 h = 1.5 j = l font = "TimesRoman"
          "Whiskers extend to minimum and maximum values.";
run;
quit;
```

It is worth taking special note of the MODE=INCLUDE option in the SYMBOL1
statement in Program 6.4. As stated in SAS Help and Documentation:

> By default, MODE=EXCLUDE, which excludes values outside the axis range
> from any calculations. If you control the range of values displayed on an axis by
> using HAXIS= and VAXIS= in the GPLOT procedure, or ORDER= in an AXIS
> definition, any data points that lie outside of the range of the axes are discarded
> before the calculations are done for interpolation lines. This has a particularly
> noticeable effect on the high-low interpolation methods, which include
> INTERPOL=HILO, INTERPOL=BOX, and INTERPOL=STD. Regression
> analysis also represents only part of the original data.

In other words, you should be very careful when explicitly specifying ranges in
SYMBOL statements, and you should make liberal use of the MODE=INCLUDE option.

Mean values could be added to the preceding plot by calculating the means for each treatment by visit and then merging them back with the seizures data. Then, a PLOT2 statement could be used in PROC GPLOT to overlay the mean values. The following is an example of a box plot modified in these ways.

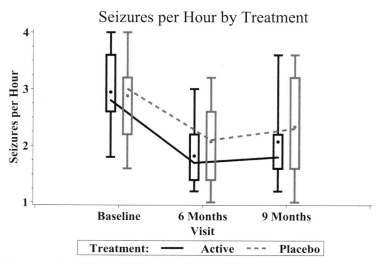

Box extends to 25th and 75th percentile. Whiskers extend to minimum and maximum values. Mean values are represented by a dot while medians are connected by the line.

Here is the SAS program that creates this box plot. The initial input DATA step is omitted because it is the same as in Program 6.4, and the changes necessary to produce this plot are highlighted.

Program 6.5 Creating a Box Plot with Means Using PROC GPLOT

```
proc format;
    value trt
        1 = "Active"
        2 = "Placebo";
run;


**** SORT THE DATA AND GET THE MEAN SEIZURE VALUE BY VISIT AND
**** TREATMENT.;
proc sort
   data = seizures;
       by visit trt;
run;


proc univariate
    data = seizures noprint;
```

```
      by visit trt;

      var seizures;
      output out = stats mean = mean;
run;

**** MERGE THE MEAN VALUES BACK IN WITH THE SEIZURE DATA.;
data seizures;
  merge seizures stats;
      by visit trt;
run;

**** CREATE PLOTVISIT VARIABLE WHICH IS A SLIGHTLY OFFSET VISIT
**** VALUE TO MAKE TRT DISTINGUISHABLE ON THE X AXIS.  OTHERWISE,
**** TREATMENT 1 AND 2 WOULD HAVE OVERLAPPING BOXES.;
data seizures;
   set seizures;

   if trt = 1 then
      plotvisit = visit - .1;
   else
      plotvisit = visit + .1;
run;

**** DEFINE GRAPHICS OPTIONS:  SET DEVICE DESTINATION TO MS
**** OFFICE CGM FILE, REPLACE ANY EXISTING CGM FILE, RESET ANY
**** SYMBOL DEFINITIONS, AND DEFINE DEFAULT FONT TYPE.;
filename fileref 'C:\whisker_mean.cgm';
goptions
   device = cgmof971
   gsfname = fileref
   gsfmode = replace
   reset = symbol
   colors = (black)
   chartype = 6;

**** SET SYMBOL DEFINITIONS.
**** SET LINE THICKNESS WITH WIDTH AND BOX WIDTH WITH BWIDTH.
**** ACTIVE DRUG IS DEFINED AS A SOLID BLACK LINE IN SYMBOL1
**** AND PLACEBO GETS A DASHED GRAY LINE IN SYMBOL2.
**** VALUE = NONE SUPPRESSES THE ACTUAL DATA POINTS.
**** BOXJT00 MEANS TO CREATE BOX PLOTS WITH BOXES (25TH AND 75TH
**** PERCENTILES - THE INTERQUARTILE RANGE) JOINED (J) AT THE
**** MEDIANS WITH WHISKERS EXTENDING TO THE MINIMUM AND MAXIMUM
**** VALUES (00) AND TOPPED/BOTTOMED (T) WITH A DASH.
**** MODE = INCLUDE OPTION ENSURES THAT VALUES THAT MIGHT FALL
**** OUTSIDE OF THE EXPLICITLY STATED AXIS ORDER WOULD BE
```

```
**** INCLUDED IN THE BOX AND WHISKER DEFINITION.;
symbol1 width = 28 bwidth = 3 color = black line = 1 value =
none
        interpol = BOXJT00 mode = include;
symbol2 width = 28  bwidth = 3 color = gray line = 2 value =
none
        interpol = BOXJT00 mode = include;
**** ADD TWO NEW SYMBOL STATEMENTS TO PLOT THE MEAN VALUES.;
symbol3 color = black value = dot;
symbol4 color = gray  value = dot;

**** DEFINE THE LEGEND FOR THE BOTTOM CENTER OF THE PAGE.;
legend1
   frame
   value = (height = 1.5)
   label = (height = 1.5 justify = right 'Treatment:' )
   position = (bottom center outside);

**** DEFINE VERTICAL AXIS OPTIONS.;
axis1 label = (h = 1.5 r = 0 a = 90 "Seizures per Hour")
      value = (h = 1.5 )
      minor = (n = 3);

**** DEFINE HORIZONTAL AXIS OPTIONS.
**** THE HORIZONTAL AXIS MUST GO FROM 0 TO 4 HERE BECAUSE OF THE
**** OFFSET APPLIED TO VISIT.  NOTICE THAT THE VALUE FOR VISIT
**** OF 0 AND 4 IS SET TO BLANK.;
axis2 label = (h = 1.5  "Visit")
      value = (h = 1.5 " " "Baseline"  "6 Months" "9 Months" "")
      order = (0 to 4 by 1)
      minor = none;

**** ADD NEW AXIS FOR PLOT2 STATEMENT BELOW.  WHITE IS USED TO
**** MAKE THE AXIS INVISIBLE ON THE PLOT.;
axis3 color = white
      label = (color = white h = .3 "   " )
      value = (color = white h = .3)
      order = (1 to 4 by 1);
```

```
**** CREATE BOX PLOT.  VISIT IS ON THE X AXIS, SEIZURES ARE ON
**** THE Y AXIS, AND THE VALUES ARE PLOTTED BY TREATMENT. THE
**** PLOT2 STATEMENT IS RESPONSIBLE FOR PLACING THE MEAN VALUES
**** ON THE PLOT.;
proc gplot
   data = seizures;

   plot seizures * plotvisit = trt  /vaxis = axis1
                                     haxis = axis2
                                     legend = legend1
                                     noframe;
   plot2 mean * plotvisit = trt /vaxis = axis3
                                    nolegend;

   format trt trt.;

   title1 h = 2 font = "TimesRoman"
        "Seizures per Hour by Treatment";
   footnote1 h = 1.5 j = l font = "TimesRoman"
        "Box extends to 25th and 75th percentile.  Whiskers"
        " extend to";
   footnote2 h = 1.5 j = l font = "TimesRoman"
        "minimum and maximum values.  Mean values are"
        " represented by";
   footnote3 h = 1.5 j = l font = "TimesRoman"
        " a dot while medians are connected by the line.";
   footnote4 h = .5 "   ";
run;
quit;
```

There are other ways to create box plots in SAS. If you want a quick box plot, you can use PROC UNIVARIATE with the PLOT option. A more elegant option is to use PROC BOXPLOT, which was introduced in SAS 8. PROC BOXPLOT is explained in detail in the *SAS/STAT User's Guide*. The following is a sample of the same box plot as in the previous example, but created with PROC BOXPLOT.

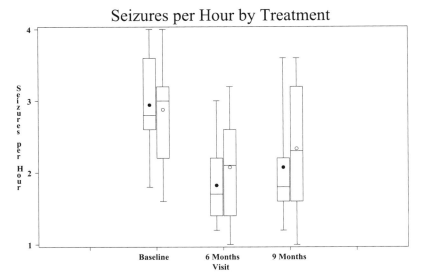

Seizures per Hour by Treatment

Box extends to 25th and 75th percentile. Whiskers extend to minimum and maximum values. Mean values are represented by a dot while medians are horizontal lines.

Here is the SAS program that creates this PROC BOXPLOT box plot. The initial input DATA step is omitted because it remains the same as in Programs 6.4 and 6.5.

Program 6.6 Creating a Box Plot with Means Using PROC BOXPLOT

```
**** CREATE PLOTVISIT VARIABLE WHICH IS A SLIGHTLY OFFSET VISIT
**** VALUE TO MAKE TRT DISTINGUISHABLE ON THE X AXIS.  OTHERWISE,
**** TREATMENT 1 AND 2 WOULD HAVE OVERLAPPING BOXES.;
data seizures;
   set seizures;

   if trt = 1 then
      plotvisit = visit - .1;
   else if trt = 2 then
      plotvisit = visit + .1;
run;


**** SORT DATA FOR PROC BOXPLOT.;
proc sort
  data = seizures;
      by plotvisit;
run;
```

```
**** FORMATS FOR BOX PLOT.;
proc format;
   value trt
      1 = "Active"
      2 = "Placebo";
   value visit
      1 = "Baseline"
      2 = "6 Months"
      3 = "9 Months"
      other = " ";
run;

**** DEFINE GRAPHICS OPTIONS:  SET DEVICE DESTINATION TO MS
**** OFFICE CGM FILE, REPLACE ANY EXISTING CGM FILE, RESET ANY
**** SYMBOL DEFINITIONS, AND DEFINE DEFAULT FONT TYPE.;
filename fileref 'C:\whisker_proc_boxplot.cgm';
goptions
   device = cgmof971
   gsfname = fileref
   gsfmode = replace
   reset = symbol
   colors = (black)
   chartype = 6;

**** SET SYMBOL DEFINITIONS.;
symbol1 value = dot;
symbol2 value = circle;

**** CREATE BOX PLOT.  VISIT IS ON THE X AXIS, SEIZURES ARE ON
**** THE Y AXIS, AND THE VALUES ARE PLOTTED BY TREATMENT.;
proc boxplot
   data = seizures;

   plot seizures * plotvisit = trt / haxis = (0,1,2,3,4)
                                     boxwidth = 3
                                     hoffset = 0;

   format trt trt. plotvisit visit.;
   label plotvisit = "Visit";

   title1 h = 2 font = "TimesRoman"
        "Seizures per Hour by Treatment";
   footnote1 h = 1.5 j = l font = "TimesRoman"
        "Box extends to 25th and 75th percentile.  Whiskers"
        " extend to";
```

```
        footnote2 h = 1.5 j = 1 font = "TimesRoman"
                "minimum and maximum values.  Mean values are"
                " represented by";
        footnote3 h = 1.5 j = 1 font = "TimesRoman"
                "a dot while medians are horizontal lines.";
        footnote4 h = .5 "   ";
    run;
    quit;
```

Note that PROC BOXPLOT uses SAS/GRAPH statements such as GOPTIONS and SYMBOL statements. There is a vast array of display options in the PROC BOXPLOT PLOT statement not used here. You may want to investigate those options if you decide to enhance the box plot further.

Creating an Odds Ratio Plot

The following is an example of an odds ratio plot. It shows the odds ratios for clinical therapy, race, gender, and baseline pain score with regard to the overall clinical success of a patient.

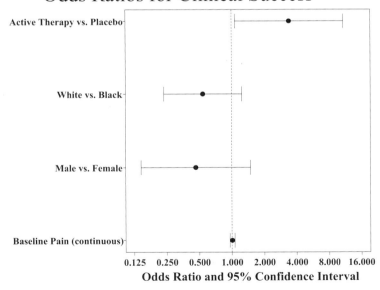

Here is the SAS program that creates the preceding graph. It is a bit complex, because SAS/GRAPH does not provide horizontal box plots and this is typically what is desired for odds ratio plots. So, this sample program relies extensively on the Annotate facility to produce the plot. Notes follow the program.

Program 6.7 Creating an Odds Ratio Plot Using PROC GPLOT

```
**** INPUT SAMPLE PAIN DATA.;
data pain;                                                    ❶
label success  = "Therapy Success"
      trt      = "Treatment"
      male     = "Male"
      race     = "Race"
      basepain = "Baseline Pain Score";
input success trt male race basepain @@;
datalines;
1 0 1 3 20    1 0 1 1 31    1 0 1 2 40    1 0 1 1 50    1 1 2 1 60
1 1 2 1 22    0 0 1 2 23    1 1 2 1 20    0 0 2 2 20    0 0 2 1 23
1 0 2 2 20    1 1 1 1 25    1 1 1 1 20    1 1 2 1 20    1 1 2 2 20
1 1 1 1 10    1 0 2 1 25    0 0 1 3 40    1 0 1 1 20    1 0 1 1 20
0 0 1 3 24    1 1 1 1 30    0 1 1 2 20    0 1 2 1 21    0 1 1 2 34
0 0 2 1 20    1 0 1 2 20    1 0 1 1 20    1 0 1 2 20    1 1 2 1 55
1 1 1 3 22    1 1 1 1 34    1 1 1 2 40    1 1 1 1 50    1 1 1 1 60
0 0 2 1 20    0 0 2 2 20    0 0 2 1 20    0 0 2 2 20    0 0 1 1 20
1 1 1 2 25    1 1 1 1 23    1 0 2 1 20    1 1 2 1 20    1 0 1 2 22
1 0 1 1 11    1 0 1 1 33    0 0 2 3 40    1 0 1 1 20    0 1 1 1 21
1 1 2 3 24    1 0 2 1 30    1 1 1 2 20    1 1 2 1 21    0 1 1 2 33
0 0 2 1 20    1 1 1 2 22    1 1 2 1 20    1 1 1 2 20    1 0 1 1 50
0 0 1 1 55    0 0 1 2 12    0 1 1 1 20    1 1 1 2 22    1 1 1 1 12
;

**** GET ADJUSTED ODDS RATIOS FROM PROC LOGISTIC AND PLACE
**** THEM IN DATA SET WALD.;
ods output CloddsWald = odds;                                 ❷
proc logistic
   data = pain
   descending;

   model success = basepain male race trt / clodds = wald;
run;
ods output close;

***** RECATEGORIZE EFFECT FOR Y AXIS FORMATTING PURPOSES.❸
data odds;
   set odds;

   select(effect);
      when("basepain") y = 1;
      when("male")     y = 2;
```

```
        when("race")      y = 3;
        when("trt")       y = 4;
        otherwise;
    end;
run;

**** FORMAT FOR EFFECT;
proc format;
    value effect
        1 = "Baseline Pain (continuous)"
        2 = "Male vs. Female"
        3 = "White vs. Black"
        4 = "Active Therapy vs. Placebo";
run;

**** ANNOTATE DATA SET TO DRAW THE HORIZONTAL LINE, ESTIMATE, AND
**** WHISKER.;
data annotate;
    set odds;

    length function $ 8 position xsys ysys $ 1;

    i = 0.10;     **** whisker width.;
    basey = y;    **** hang onto row position.;

    **** set coordinate system and positioning.;
    position = '5';
    xsys = '2';
    ysys = '2';
    line = 1;

    **** plot estimates on right part of the graph.;
    if oddsratioest ne . then
        do;
            *** place a DOT at OR estimate.;
            function = 'SYMBOL';
            text = 'DOT';
            size = 1.5;
            x = oddsratioest;
            output;
            *** move to LCL.;
            function = 'MOVE';
            x = lowercl;
            output;
            *** draw line to UCL.;
            function = 'DRAW';
            x = uppercl;
```

```
            output;
            *** move to LCL bottom of tick mark.;
            function = 'MOVE';
            x = lowercl;
            y  = basey - i;
            output;
            *** draw line to top of tick mark.;
            function = 'DRAW';
            x = lowercl;
            y = basey + i;
            output;
            *** move to UCL bottom of tick mark.;
            function = 'MOVE';
            x = uppercl;
            y = basey - i;
            *** draw line to top of tick mark.;
            output;
            function = 'DRAW';
            x = uppercl;
            y = basey + i;
            output;
        end;
run;


**** DEFINE GRAPHICS OPTIONS:  SET DEVICE DESTINATION TO MS
**** OFFICE CGM FILE, REPLACE ANY EXISTING CGM FILE, RESET ANY
**** SYMBOL DEFINITIONS, AND DEFINE DEFAULT FONT TYPE.;
filename fileref 'C:\odds_ratio.cgm';
goptions
    device = cgmof971
    gsfname = fileref
    gsfmode = replace
    reset = symbol
    colors = (black)
    chartype = 6;


**** DEFINE SYMBOL TO MAKE THE GRAPH SPACE THE PROPER SIZE BUT
**** NOT TO ACTUALLY PLOT ANYTHING.;
symbol color = black interpol = none value = none repeat = 2;


**** DEFINE HORIZONTAL AXIS OPTIONS.;
axis1 label = (h = 1.5 "Odds Ratio and 95% Confidence Interval")
      value = (h = 1.2)
      logbase = 2
      logstyle = expand
      order = (0.125,0.25,0.5,1,2,4,8,16)
      offset = (2,2) ;
```

```
**** DEFINE VERTICAL AXIS OPTIONS.;
axis2 label = none
      value = (h = 1.2)
      order = (1 to 4 by 1)
      minor = none
      offset = (2,2);

**** CREATE THE ODDS RATIO PLOT.  THIS IS DONE PRIMARILY THROUGH
**** THE INFORMATION IN THE ANNOTATION DATA SET. PUT A HORIZONTAL
**** REFERENCE LINE AT 1 WHICH IS THE LINE OF SIGNIFICANCE.;
proc gplot
   data = odds;

   plot y * lowercl y * uppercl / anno = annotate
                                  overlay
                                  href = 1
                                  lhref = 2
                                  haxis = axis1
                                  vaxis = axis2;
   format y effect.;
   title1 h = 2.5 font = "TimesRoman"
          "Odds Ratios for Clinical Success";
run;
quit;
```

Notes for Program 6.7:

❶ This is the sample pain data set with a dependent variable called "success," which is 1 if the patient achieved clinical success and 0 otherwise. The treatment, gender, race, and baseline pain scores serve as the independent variables.

❷ Here we use PROC LOGISTIC to obtain our adjusted odds ratios and 95% confidence limits. ODS is used to send the statistics to the "odds" data set. Note that the DESCENDING option is used in PROC LOGISTIC to model for the probability that success = 1.

❸ In order to make the Annotate data set work well with the Y axis, the "effect" variable in the odds data set is translated to a numeric value. The subsequent PROC FORMAT creates the format used in the PROC GPLOT.

Note that *hazard ratios* can be plotted in the same way that odds ratios are plotted. Hazard ratios are created using the *Cox proportional hazards model* through PROC PHREG.

Creating a Kaplan-Meier Survival Estimates Plot

The following is an example of a Kaplan-Meier survival estimates plot. In this plot, we are comparing the time to death for three different treatment regimens. The Kaplan-Meier survival estimate is on the Y axis, and time is represented on the X axis. Each step in the graph lines represents an event.

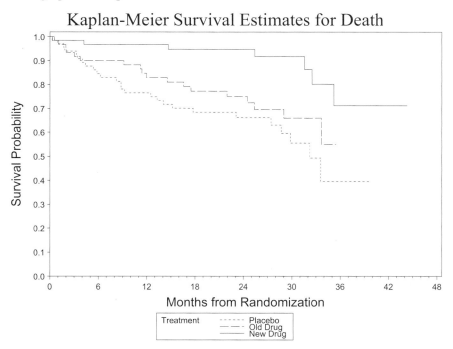

Here is the SAS program that creates this Kaplan-Meier estimates plot.

Program 6.8 Creating a Kaplan-Meier Estimates Plot Using PROC GPLOT

```
****  INPUT SAMPLE TIME TO DEATH DATA.
****  DAYS TO DEATH IS THE NUMBER OF DAYS FROM RANDOMIZATION TO
****  DEATH OR LAST PATIENT FOLLOW-UP.  DEATHCENSOR VARIABLE IS A
****  "1" IF THE PATIENT DIED AT THE TIME TO EVENT AND IT IS A "0"
****  IF THE PATIENT WAS ALIVE AT LAST FOLLOW-UP.  TO SAVE SPACE,
****  PATIENT ID HAS BEEN OMITTED FROM THIS SAMPLE DATA SET.;
data death;
label trt         = "Treatment"
      daystodeath = "Days to Death"
      deathcensor = "Death Censor";
```

```
input trt $ daystodeath deathcensor @@;
datalines;
A   52    1     A   825   0     C   693   0     C   981   0
B   279   1     B   826   0     B   531   0     B   15    0
C   1057  0     C   793   0     B   1048  0     A   925   0
C   470   0     A   251   1     C   830   0     B   668   1
B   350   0     B   746   0     A   122   1     B   825   0
A   163   1     C   735   0     B   699   0     B   771   1
C   889   0     C   932   0     C   773   1     C   767   0
A   155   0     A   708   0     A   547   0     A   462   1
B   114   1     B   704   0     C   1044  0     A   702   1
A   816   0     A   100   1     C   953   0     C   632   0
C   959   0     C   675   0     C   960   1     A   51    0
B   33    1     B   645   0     A   56    1     A   980   1
C   150   0     A   638   0     B   905   0     B   341   1
B   686   0     B   638   0     A   872   1     C   1347  0
A   659   0     A   133   1     C   360   0     A   907   1
C   70    0     A   592   0     B   112   0     B   882   1
A   1007  0     C   594   0     C   7     0     B   361   0
B   964   0     C   582   0     B   1024  1     A   540   1
C   962   0     B   282   0     C   873   0     C   1294  0
B   961   0     C   521   0     A   268   1     A   657   0
C   1000  0     B   9     1     A   678   0     C   989   1
A   910   0     C   1107  0     C   1071  1     A   971   0
C   89    0     A   1111  0     C   701   0     B   364   1
B   442   1     B   92    1     B   1079  0     A   93    0
B   532   1     A   1062  0     A   903   0     C   792   0
C   136   0     C   154   0     C   845   0     B   52    0
A   839   0     B   1076  0     A   834   1     A   589   0
A   815   0     A   1037  0     B   832   0     C   1120  0
C   803   0     C   16    1     A   630   0     B   546   0
A   28    1     A   1004  0     B   1020  0     A   75    0
C   1299  0     B   79    0     C   170   0     B   945   0
B   1056  0     B   947   0     A   1015  0     A   190   1
B   1026  0     C   128   1     B   940   0     C   1270  0
A   1022  1     A   915   0     A   427   1     A   177   1
C   127   0     B   745   1     C   834   0     B   752   0
A   1209  0     C   154   0     B   723   0     C   1244  0
C   5     0     A   833   0     A   705   0     B   49    0
B   954   0     B   60    1     C   705   0     A   528   0
A   952   0     C   776   0     B   680   0     C   88    0
C   23    0     B   776   0     A   667   0     C   155   0
B   946   0     A   752   0     C   1076  0     A   380   1
B   945   0     C   722   0     A   630   0     B   61    1
C   931   0     B   2     0     B   583   0     A   282   1
A   103   1     C   1036  0     C   599   0     C   17    0
C   910   0     A   760   0     B   563   0     B   347   1
```

```
B  907  0    B  896  0    A  544  0    A  404  1
A    8  1    A  895  0    C  525  0    C  740  0
C   11  0    C  446  1    C  522  0    C  254  0
A  868  0    B  774  0    A  500  0    A   27  0
B  842  0    A  268  1    B  505  0    B  505  1
;
run;

**** PERFORM LIFETEST AND EXPORT SURVIVAL ESTIMATES.;
ods output ProductLimitEstimates = survest;
proc lifetest
   data = death;

   time daystodeath * deathcensor(0);
   strata trt;
run;
ods output close;

proc sort
   data = survest;
      by trt daystodeath;
run;

**** PREPARE THE SURVIVAL ESTIMATE DATA FOR PLOTTING.;
data survest;
   set survest;
      by trt;

      **** CALCULATE MONTH FOR PLOTTING.;
      month = (daystodeath / 30.417);   *** = 365/12;

      **** ENSURE THAT THE LAST TIME TO EVENT VALUE WITHIN A
      **** TREATMENT IS REPRESENTED ON THE PLOT IN THE CASE
      **** THAT IT WAS NOT A DEATH.;
      retain lastsurv;
      if first.trt then
         lastsurv = .;

      if survival ne . then
         lastsurv = survival;

      if last.trt and survival = . then
         survival = lastsurv;

      **** REMOVE RECORDS WHERE ESTIMATE MISSING;
      if survival ne . or last.trt;
run;
```

```
**** DEFINE GRAPHICS OPTIONS:  SET DEVICE DESTINATION TO MS
**** OFFICE CGM FILE, REPLACE ANY EXISTING CGM FILE, RESET ANY
**** SYMBOL DEFINITIONS, AND DEFINE DEFAULT FONT TYPE.;
filename fileref 'C:\survival.cgm';
goptions
    device = cgmof971
    gsfname = fileref
    gsfmode = replace
    reset = symbol
    colors = (black);


**** DEFINE SYMBOLS.  STEPJL INDICATES TO CREATE A STEP
**** FUNCTION AND TO JOIN THE POINTS AT THE LEFT.;
symbol1 c = black line = 2 v = NONE interpol = stepjl ;
symbol2 c = black line = 4 v = NONE interpol = stepjl ;
symbol3 c = black line = 1 v = NONE interpol = stepjl ;


**** DEFINE HORIZONTAL AXIS OPTIONS.;
axis1   order = (0 to 48 by 6)
        label = (h = 1.5 "Months from Randomization")
        minor = (number = 5)
        offset = (0,1);


**** DEFINE VERTICAL AXIS OPTIONS.;
axis2   order = (0 to 1 by .1)
        label = (h = 1.5 angle = 90 "Survival Probability")
        minor = none
        offset = (0,.75);


**** DEFINE THE LEGEND FOR THE BOTTOM CENTER OF THE PAGE.;
legend1 label = ("Treatment")
        order = ("A" "B" "C")
        value = ("Placebo" "Old Drug" "New Drug")
        down = 3
        position = (bottom center)
        frame;


**** CREATE KM PLOT.  SURVIVAL ESTIMATE IS ON THE Y AXIS, MONTHS
**** ARE ON THE X AXIS, AND THE VALUES ARE PLOTTED BY TREATMENT.;
proc gplot
    data = survest;

    plot survival * month = trt / haxis = axis1
                                  vaxis = axis2
                                  legend = legend1 ;
```

```
    format survival 4.1 month 2.;
    title1 h = 2 font = "TimesRoman"
          "Kaplan-Meier Survival Estimates for Death";
run ;
quit ;
```

Creating a Failure Estimate Plot

On occasion it is necessary to produce "failure" estimate plots instead of "survival" estimates plots. Fortunately, this requires only a simple modification to the preceding Kaplan-Meier survival estimates program. The only changes necessary to this program to get a failure plot are to alter the title and axis labels, and to change the "survival" variable reference to "failure" because the "failure" variable is also present in the ProductLimitEstimates data set. The resulting failure estimate plot looks like the following:

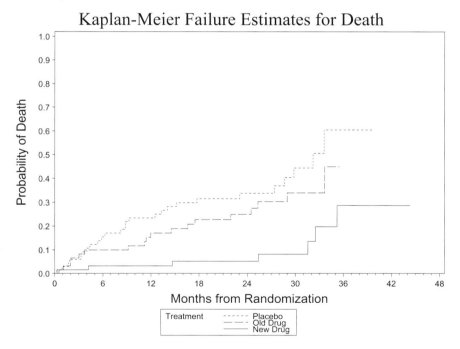

Creating a Survival Estimates Plot Directly from PROC LIFETEST

Another option for producing Kaplan-Meier survival estimates plots is to use the built-in PROC LIFETEST high-resolution plots. These plots allow you to avoid using SAS/GRAPH and PROC GPLOT to produce the survival estimates plots. For instance, Program 6.9 creates the preceding survival estimates plot directly from PROC LIFETEST. The initial input DATA step has been omitted because it is the same as in Program 6.8.

The following graph is produced by the code in Program 6.9.

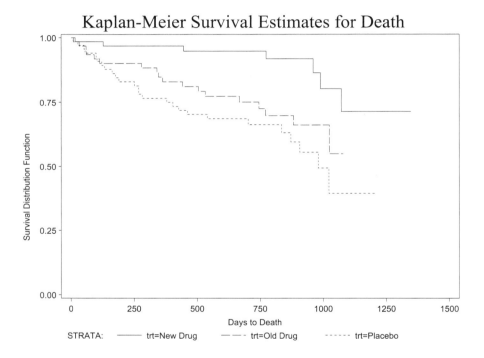

Program 6.9 Creating a Kaplan-Meier Survival Estimates Plot Using PROC LIFETEST

```
proc format;
   value $trt
      "A" = "Placebo"
      "B" = "Old Drug"
      "C" = "New Drug";
run;

**** DEFINE GRAPHICS OPTIONS:  SET DEVICE DESTINATION TO MS
**** OFFICE CGM FILE, REPLACE ANY EXISTING CGM FILE, RESET ANY
**** SYMBOL DEFINITIONS, AND DEFINE DEFAULT FONT TYPE.;
filename fileref "C:\survival_from_lifetest.cgm";
goptions
   device = cgmof971
   gsfname = fileref
   gsfmode = replace
   reset = symbol
   colors = (black);

**** DEFINE SYMBOLS.;
symbol1 c = black line = 1 v = NONE;
symbol2 c = black line = 4 v = NONE;
symbol3 c = black line = 2 v = NONE;


**** CREATE KAPLAN-MEIER PLOT WITH PROC LIFETEST.;
proc lifetest
   data = death
   plots = (s)
   censoredsymbol = none
   eventsymbol = none;

   time daystodeath * deathcensor(0);
   strata trt;

   format trt $trt.;

   title1 h = 2 font = "TimesRoman"
          "Kaplan-Meier Survival Estimates for Death";
run;
```

The Kaplan-Meier survival estimates plots are instantiated by specifying "PLOTS = (S)" in the PROC LIFETEST statement. To show just the line itself, "CENSOREDSYMBOL = NONE" is specified to hide the censored observations in the plot. "EVENTSYMBOL = NONE" is specified here to hide the event points, although this is the default setting for

PROC LIFETEST. Note how the GOPTIONS and SYMBOL statements are used by the PROC LIFETEST plots. However, AXIS statements and LEGEND statements are not used by PROC LIFETEST plots.

If you can accept the axis and legend definitions that are provided by the survival estimates plots from PROC LIFETEST, then this is a faster way to produce your plots than the PROG GPLOT method previously shown. It is only when you need precise control of your plot appearance that PROG GPLOT is necessary.

Output Options

Although all of the examples in this chapter export graphs to Microsoft Windows CGM files, SAS/GRAPH has the ability to export graphs to many different destinations. The two basic ways to export SAS/GRAPH output are with SAS graphics *drivers* and with an ODS destination specification.

Selecting Graphics Drivers

The examples in this book are created with SAS graphics drivers via the GOPTIONS DEVICE= specification. This is done because the resulting output is to be included in a Microsoft Word document and the CGM drivers supplied by SAS provide the most attractive output for Microsoft Word files. However, there are many other devices that can be included in the DEVICE= specification. To see a list of your available devices, simply type "PROC GDEVICE; RUN;" in an interactive SAS program window and submit it. A full directory of available graphics devices will be provided to you. Many of the devices are specific printer drivers, but many generic graphics drivers are available as well. PostScript, CGM, HTML, Bitmap, EMF, GIF, JPEG, TIFF, PNG, and WMF are some of the generic print drivers that can be selected.

To change the graphics file you want to produce, you only need to make a minor adjustment to the FILENAME and GOPTIONS statements. Just change the shaded text below in your SAS/GRAPH code to the appropriate settings:

```
filename fileref "C:\your_filename_here.valid_extension";

goptions
   device = DEVICE_NAME_FROM_PROC_GDEVICE
   gsfname = fileref;
```

Using ODS Destinations for SAS/GRAPH

Another option for exporting SAS graphics is to employ ODS to send the graphics to various destinations. This is fairly easy to do, as it requires no knowledge of the SAS/GRAPH graphics drivers. For example, if you wanted to use the previous survival estimate SAS/GRAPH graphs to create RTF output for Word, PDF output for Acrobat, and HTML output for Internet Explorer, you could simply add some ODS statements like these:

```
ods rtf;  **** SEND OUTPUT TO RTF DESTINATION;
ods pdf;  **** SEND OUTPUT TO PDF DESTINATION;
ods html; **** SEND OUTPUT TO HTML DESTINATION;
proc gplot
   data = plotset;

   plot survival * month = trt ;

   title1 h = 2 "Kaplan-Meier Survival Estimates";
run;
quit ;
ods rtf close;
ods pdf close;
ods html close;
```

This exports your survival plots to the three desired destinations. The results look like the following three graphs:

RTF output:

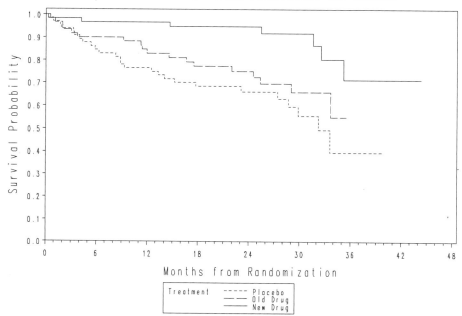

PDF output:

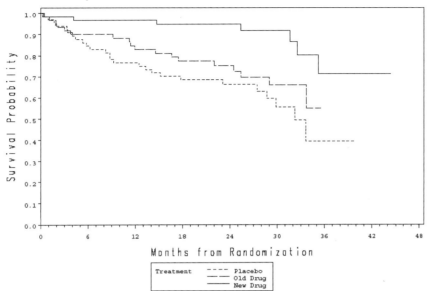

HTML output:

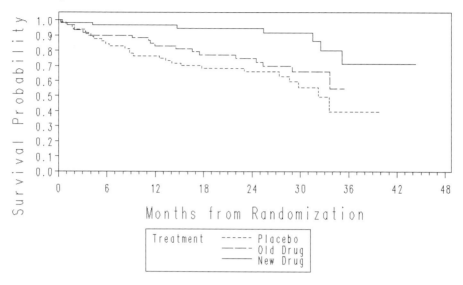

Although ODS destinations are easy to use with SAS/GRAPH, the appearance of the results may vary considerably. A graph produced by the GOPTIONS DEVICE = CGMOF97L statement and a graph created by ODS RTF often look different, even though they are both likely to be destined for Microsoft Word. The best advice is to experiment with various graphics output destinations and use the one that best suits your needs.

Using SAS/GRAPH Assistants

SAS/GRAPH can have a rather large learning curve. Once you have used it extensively, you can produce any graph you can imagine, but initially the software can be daunting. For this reason, there are some assisting technologies within SAS that can help get you started creating graphs.

Graph-N-Go

Part of the SAS windowing environment is a graph-building tool called Graph-N-Go. You can access this tool from the SAS toolbar by selecting "Solutions," "Reporting," and then "Graph-N-Go." This invokes the Graph-N-Go facility. It lets you select a data set and a graph type, and then choose which graph you want to create with a point-and-click interface.

SAS Enterprise Guide

Because SAS Enterprise Guide 3.0 is part of Base SAS (beginning with SAS 9.1), you might consider using it to get started with SAS graphics. SAS Enterprise Guide 3.0 has a well-designed interface for building SAS graphs. Simply select a data set from the Process Flow window and click "Graph" on the toolbar. You can then use a point-and-click interface to build many different kinds of graphs.

ODS Graphics

ODS Graphics, an experimental tool in SAS 9.1, may be another option to consider when producing graphics output with SAS. Because ODS Graphics is experimental, adequate testing should be done before including ODS Graphics in production reports. For the future, ODS Graphics bears watching as it matures. The idea behind ODS Graphics is to be able to turn a simple switch within your SAS code to produce standard SAS graphs. For example, the shaded statements in the following SAS code send all standard graphs from PROC LOGISTIC to a file called "logistic.pdf":

```
ods pdf file = 'C:\logistic.pdf';
ods graphics;
proc logistic
    data = pain
    descending;

    model success = basepain gender race trt;
graphics all;
run;
ods graphics off;
ods pdf close;
```

There is detailed information about ODS Graphics in the SAS 9.1 documentation. ODS Graphics does allow you to avoid the complexities of learning SAS/GRAPH, but if you want to alter the graphics created with ODS Graphics, you must learn the complexities of the ODS Graphics template language. So if you want to modify ODS Graphics output, you are trading off learning one set of complexities for another.

These SAS graphics technologies can be used as a starting point for making the graphs you want. For instance, say you have not had to create a bar chart in SAS for years. In that case it really helps to invoke SAS Enterprise Guide or Graph-N-Go to build your graph. Chances are that these tools will get you about 90% of the way to your desired graph. Then, if necessary, you can export the SAS program being built by the SAS tool for final SAS/GRAPH touch-up work.

When You Should Use SAS/GRAPH

When should you use SAS/GRAPH? That is a question that deserves some attention. SAS/GRAPH is a wonderful choice for graphics if the following are true:

- You need to reproduce graphics periodically over time.
- You need to produce more than a few of a similar type of graph.
- You have all of your other analysis and reporting programming in SAS.

If you need to reproduce graphics periodically over time, then your stored SAS/GRAPH programs can be rerun whenever you want. Perhaps you have four sequential data safety and monitoring reports and then a subsequent final study report to produce for a clinical trial. In this case, once the SAS/GRAPH code has been set up, it can be run on the five occasions when it is needed. For another example, perhaps you have to produce 300 laboratory data plots. In this case SAS/GRAPH makes a great choice, because you can use the power of BY-statement processing and/or the SAS macro language to turn one PROC GPLOT into 300 plots. Finally, if you are doing all of your reporting in SAS, then completing the graphical components in SAS only makes sense.

If, however, you have a simple graph to produce, and there is only one of them, and you only ever need to make the graph once, then a customized SAS/GRAPH may be more effort than it is worth. In this case I would recommend using one of the assisting SAS graphics technologies listed in the previous section. In the event that you need to put multiple plots on a single page, you can use PROC GREPLAY. (A good resource for this is the SAS Press book *Multiple-Plot Displays: Simplified with Macros,* by Perry Watts.)

Chapter 7

Performing Common Analyses and Obtaining Statistics

This chapter briefly discusses many common types of clinical trial data analyses and where in SAS you can obtain the statistics you need to present in your tables. Because it is usually up to the clinical trial statistician to design the inferential testing and data modeling that are required, this book does not go into depth on statistical methods. For more detail on statistical methodology, you can begin by referring to the SAS Press books *Common Statistical Methods for Clinical Research with SAS Examples*, by Glenn Walker, and *Analysis of Clinical Trials Using SAS: A Practical Guide,* by Alex Dmitrienko et al. This chapter provides some practical examples of how to extract statistics from SAS procedures for presentation in your summary tables and graphs.

Obtaining Descriptive Statistics

Descriptive statistics typically make up the majority of clinical trial reporting. You use descriptive statistics to characterize a data distribution without making inferences about the data. You can provide descriptive statistics on categorical and continuous data. Categorical data are data that are divided into a set of definite responses such as gender, race, and ethnic class. Continuous data are data that can have any value within a range of acceptable limits, such as age, weight, and height. Two primary sources for obtaining descriptive statistics in SAS are PROC FREQ for categorical data and PROC UNIVARIATE for continuous data.

Using PROC FREQ to Export Descriptive Statistics

PROC FREQ can be used to export frequencies of categorical data simply by specifying the OUT= option in the TABLES statement like this:

```
proc freq
   data = demog;
      by trt;

      tables a * b /out = freqs;
run;
```

The resulting "freqs" data set contains the BY variable "trt," "a," "b," and the cell frequency "count" and percentage "percent" variables. Row and column percentages can be added to the output data set by specifying the OUTPCT option. If you also want the "totals" row and column that you see in your PROC FREQ listing output, you can use ODS to export that to a data set called "freqs":

```
ods output CrossTabFreqs = freqs;
proc freq
   data = demog;
      by trt;

      tables a * b;
run;
ods output close;
```

Remember that if you do not want to exclude missing values from your counts or from the denominators of your percentage calculations, you need to specify the MISSING option in your TABLES statement.

Using PROC UNIVARIATE to Export Descriptive Statistics

PROC UNIVARIATE can be used to export a large number of descriptive statistics on continuous variables simply by specifying the OUTPUT statement like this:

```
proc univariate
   data = demog;
      by trt;

      var x;
      output out = dsetname variables;
run;
```

In this case "dsetname" is the name of your output data set, and "*variables*" is a variable name list of one or more statistics in the following table.

Variable Name	Variable Description*
CSS	Corrected sum of squares
CV	Coefficient of variation
KURTOSIS	Kurtosis
MAX	Largest value
MEAN	Sample mean
MIN	Smallest value
MODE	Most frequent value
N	Sample size
NMISS	Number of missing values
NOBS	Number of observations
RANGE	Range
SKEWNESS	Skewness
STD	Standard deviation
STDMEAN	Standard error of the mean
SUM	Sum of the observations
SUMWGT	Sum of the weights
USS	Uncorrected sum of squares
VAR	Variance
P1	1st percentile
P5	5th percentile
P10	10th percentile
Q1	Lower quartile (25th percentile)
MEDIAN	Median (50th percentile)
Q3	Upper quartile (75th percentile)
P90	90th percentile
P95	95th percentile
P99	99th percentile
QRANGE	Interquartile range (Q3–Q1)

* From SAS OnlineDoc 9.1.3.

PROC MEANS, PROC SUMMARY, and PROC TABULATE are other SAS procedures that you can use to get descriptive statistics and place them into output data sets. However, those procedures do not offer any descriptive statistical variables that you cannot get from PROC FREQ or PROC UNIVARIATE.

Obtaining Inferential Statistics from Categorical Data Analysis

For categorical variables you may be required to provide inferential statistics along with the descriptive frequency and percentage statistics. These inferential statistics are generally tests of association to determine if one categorical variable, such as treatment group, can be said to be associated with another categorical variable. This section looks at some common categorical inferential tests and shows how to get the statistics you need out of SAS. For detailed information about categorical data analysis you may want to refer to the SAS Press book *Categorical Data Analysis Using the SAS System,* by Maura Stokes et al.

Performing a 2x2 Test for Association

A very common analysis in clinical trials involves the analysis of two *binomial variables* to see if there is a *statistically significant* association between them. A binomial variable is one that can have only one of two values. For example, let's assume that we have a variable called "treatment" whose value is either a 1 to indicate active drug therapy or a 0 to indicate placebo. We also have a variable called "headache" whose value is a 1 if the patient experiences headache after therapy and a 0 if not. What we want to know is whether a change in the level of therapy is significantly associated with a change in the level of headache. The *2x2 table* looks like this:

	Active Drug	**Placebo**
Headache	?	?
No Headache	?	?

If the normal approximation to the binomial distribution is valid (that is, not more than 20% of expected cell counts are less than 5) for drug therapy and symptom of headache, then you can use the *Pearson chi-square test* to test for a difference in proportions. To get the Pearson chi-square p-value for the preceding 2x2 table, you run SAS code like the following:

```
proc freq;
   table headache * treatment / chisq;
   output out = pvalue pchi;
run;
```

The output data set "pvalue" includes a variable called "P_PCHI" that contains the Pearson chi-square *p*-value you need.

If the normal approximation to the binomial distribution is not valid (that is, more than 20% of expected cell counts are less than 5) for drug therapy and symptom of headache, then you can use *Fisher's exact test,* which is a *nonparametric test,* to test for a difference in proportions. To get the *p*-value using Fisher's exact test, you run the following SAS code:

```
proc freq;
   table headache * treatment / exact;
   output out = pvalue exact;
run;
```

The output data set "pvalue" includes a variable called "XP2_FISH" that contains the Fisher's exact *p*-value you need.

Performing an NxP Test for Association

Sometimes you have more than two levels of groups or responses and you want to test for association. In this case you have an *NxP table* that looks like this:

	Therapy 1	Therapy 2	...
Response 1	?	?	?
Response 2	?	?	?
Response 3	?	?	?
...	?	?	?

Here you can still use the Pearson chi-square test as shown in the 2x2 table example as long as your response variable is *nominal* and merely descriptive. If your response variable is *ordinal,* meaning that it is an ordered sequence, and you can use a *parametric test,* then you should use the *Mantel-Haenszel test* statistic for parametric tests of association. For instance, if in our previous example the variable called "headache" was coded as a "2" when the patient experienced extreme headache, a "1" if mild headache, and a "0" if no headache, then "headache" would be an ordinal variable. You can get the Mantel-Haenszel *p*-value by running the following SAS code:

```
proc freq;
   table headache * treatment / chisq;
   output out = pvalue pchi;
run;
```

Your Mantel-Haenszel *p*-value is found in the "P_MHCHI" variable in the "pvalue" data set.

Performing a Stratified NxP Test for Association

On occasion you need to perform a test for association between two categorical variables while *stratifying*, or controlling, for a third variable. The *Cochran-Mantel-Haenszel test* for association stratifies by a third variable to give proper weight to strata size. In the previous example, let's assume that you want to stratify your analysis by center to control for differences in center size. You can then run a Cochran-Mantel-Haenszel test like this:

```
proc freq;
   table center * treatment * headache / cmh;
   output out = pvalue cmh;
run;
```

The Cochran-Mantel-Haenszel (CMH) test contains three different tests for association. The following table describes each test and shows where the *p*-value can be found in your "pvalue" data set.

CMH Test	Description	*p*-value Variable
Nonzero Correlation	A significant *p*-value here indicates that there is a linear correlation between the two ordinal variables for at least one stratum.	P_CMHCOR
Row Mean Scores Differ	For an ordinal column variable, a significant *p*-value here indicates that the mean CMH score differs across columns for at least one stratum.	P_CMHRMS
General Association	A significant *p*-value here indicates that there is an association between the two variables for at least one stratum.	P_CMHGA

For Cochran-Mantel-Haenszel tests in PROC FREQ, you usually want to structure your TABLE statement in the following way:

```
table STRATA * TREATMENT * RESPONSE / cmh;
```

Make sure your stratification variable appears first and your ordinal response variable appears last so that your stratification is properly applied and so that your row mean score test will be meaningful.

Performing Logistic Regression

Logistic regression analysis involves trying to determine or model the result of a dependent binomial variable based on a set of independent *predictor variables*. For example, let's say that you are trying to determine if drug therapy (active drug or placebo), patient age, and history of head trauma have an effect in predicting whether a patient will be cured of headache (0=No headache, 1=Headache). You can test this model with logistic regression as follows:

```
ods output OddsRatios = odds;
ods output ParameterEstimates = estimates;

proc logistic
   descending;

   model headache = treatment age head_trauma;
run;

ods output close;
```

There are a few important things to note about this SAS code. First, the DESCENDING option is used in PROC LOGISTIC to model the probability of having a headache (headache = 1). By default, PROC LOGISTIC models to the lower value of your dependent variable. Because most event variables that you encounter in clinical trials are coded as 1 = "Yes Event" and 0 = "No Event," you generally want to use the DESCENDING option with PROC LOGISTIC. Second, ODS OUTPUT statements are used to send the covariate-adjusted odds ratios and confidence intervals to the "odds" data set and to send the parameter estimates and *p*-values to the "estimates" data set.

Obtaining Inferential Statistics from Continuous Data Analysis

For continuous variables you may be required to provide inferential statistics along with the descriptive statistics that you generate from PROC UNIVARIATE. The inferential statistics discussed here are all focused on *two-sided tests* of mean values and whether they differ significantly in either direction from a specified value or another population mean. Many of these tests of the mean are parametric tests that assume the variable being tested is *normally distributed*. Because this is often not the case with clinical trial data, we discuss substitute nonparametric tests of the population means as well. Here are some common continuous variable inferential tests and how to get the inferential statistics you need out of SAS.

Performing a One-Sample Test of the Mean

In clinical trial analyses you may want to test the mean for a single population to determine if that value differs from a hypothesized value. For example, let's say that you have the lab test value LDL and you want to know if the change-from-baseline value is significantly different from zero. There are several ways to perform this test in SAS. If you assume the change from baseline for LDL, "ldl_change," is normally distributed, you can run a *one-sample t-test* in SAS like this:

```
ods output TTests = pvalue;

proc ttest h0 = 0;
   var ldl_change;
run;

ods output close;
```

The *p*-value for the *t*-test can be found in the "Probt" variable in the "pvalue" data set.

PROC UNIVARIATE can also be used to perform the one-sample *t*-test as follows:

```
ods output TestsForLocation = pvalue;

proc univariate mu0 = 0;
   var ldl_change;
run;

ods output close;
```

The *p*-value for the *t*-test is found in the "pValue" variable in the "pvalue" data set, where the variable "test" is equal to "Student's t."

If you do not have the change-from-baseline variable, "ldl_change," calculated but you have the baseline variable, "ldl_base," and post-treatment variable, "ldl_post_trt," then you can run this comparison as a *paired t-test,* like this:

```
ods output TTests = pvalue;

proc ttest;
   paired ldl_base * ldl_post_trt;
run;

ods output close;
```

The *p*-value for the *t*-test can be found in the "Probt" variable in the "pvalue" data set.

Finally, if the "ldl_change" variable is not normally distributed, then you can run a nonparametric test on the change-from-baseline LDL value, like this:

```
ods output TestsForLocation = pvalue;

proc univariate mu0 = 0;
   var ldl_change;
run;

ods output close;
```

The *p*-value for the *sign test* or *Wilcoxon signed rank test* can be found in the "pValue" variable in the "pvalue" data set. If the variable is from a symmetric distribution, you can get the *p*-value from the Wilcoxon signed rank test, where the "Test" variable in the "pvalue" data set is "Signed Rank." If the variable is from a *skewed distribution,* you can get the *p*-value from the sign test, where the "Test" variable in the "pvalue" data set is "Sign."

Performing a Two-Sample Test of the Means

In clinical trial analyses you may want to test the means of two independent populations to determine if they are significantly different. For example, let's say that you have the lab test value LDL and you want to know if the change-from-baseline value is significantly different after treatment between active drug and placebo. If you assume the change from baseline for LDL, "ldl_change," is normally distributed, you can run a *two-sample t-test* in SAS like this:

```
ods output Equality = variance_test;
ods output TTests = pvalue;

proc ttest
   class treatment;
   var ldl_change;
run;

ods output close;

**** CHECK VARIANCES AND SELECT PROPER P-VALUE;
data pvalue;
   if _n_ = 1 then
      set variance_test(keep = probf);
   set pvalue(keep = variances probt);

   keep probt;

   if (probf <= .05 and variances = "Unequal") or
      (probf > .05 and variances = "Equal");
run;
```

This PROC TTEST runs a two-sample *t*-test to compare the LDL change-from-baseline means for active drug and placebo. ODS OUTPUT is used to send the *p*-values to a data set called "pvalue" and to send the test of equal mean variances to a data set called "variance_test." The final "pvalue" DATA step checks the *test for unequal variances*. If the test for unequal variances is significant at the alpha = .05 level, then the mean variances are unequal and the unequal variances *p*-value is kept. If the test for unequal variances is insignificant, then the equal variances *p*-value is kept. The final "pvalue" data set contains the "Probt" variable, which is the *p*-value you want.

If the two sample populations are not normally distributed, then you can use the nonparametric *Wilcoxon rank sum test* to compare the population means. The following SAS code compares the "ldl_change" change-from-baseline means for active drug and placebo:

```
proc npar1way
    wilcoxon;
    class treatment;
    var ldl_change;
    output out = pvalue wilcoxon;
run;
```

The output data set "pvalue" contains numerous Wilcoxon test statistics. Assuming that you want the two-sided normal approximation test *p*-value, the variable in the "pvalue" data set that you want is called "P2_WIL."

Performing an N-Sample Test of the Means

If there are more than two treatment groups in your clinical trial, then you may want to compare population means across all treatment groups. In this case you need to perform a *one-way analysis of variance*. Let's assume that you have the three treatment groups "Old Drug," "New Drug," and "Placebo," and you want to see if there is a statistically significant difference in LDL change from baseline between treatments. You can run the following PROC GLM to compare the treatment means:

```
proc glm
    outstat = pvalue;
    class treatment;
    model ldl_change = treatment;
run;
quit;
```

The OUTSTAT= output data set "pvalue" contains the *p*-value in the "PROB" variable. If you have multiple predictor variables, you need to use the PROC GLM ODS data set "OverallANOVA" to get the overall model *p*-value from the "ProbF" variable. These output data sets contain other variables, such as the degrees of freedom, sum of squares, mean square, and F statistic, if you need them for an ANOVA table presentation.

If the populations to be compared across treatments are not normally distributed, you can use the nonparametric *Kruskal-Wallis test* of the distributions by running PROC NPAR1WAY as follows:

```
proc npar1way
    wilcoxon;
    class treatment;
    var ldl_change;
    output out = pvalue wilcoxon;
run;
```

The Kruskal-Wallis *p*-value variable in the "pvalue" data set is called "P_KW."

Obtaining Time-to-Event Analysis Statistics

Time-to-event analysis in clinical trials is concerned with comparing the distributions of time to some event for various treatment regimens. The two nonparametric tests used to compare distributions are the *log-rank test* and the Cox proportional hazards model. The Cox proportional hazards model is more useful when you need to adjust your model for covariates.

Let's assume that you are comparing the two treatment groups of "Active Drug" and "Placebo" to see if they display different distributions for time to death. You can run this analysis with PROC LIFETEST as follows:

```
ods output HomTests = pvalue;

proc lifetest
    time daystodeath * deathcensor(0);
    strata treatment;
run;

ods output close;
```

Here we assume that the "daystodeath" variable is the number of days to death or last known date alive for the patient. The "deathcensor" variable value is 1 if the patient died and 0 if the patient did not die. The *p*-value variable for the log-rank test is called "ProbChiSq" in the "pvalue" data set where the "Test" variable equals "Log-Rank."

Now let's assume that you want to adjust for gender as a covariate in your comparison of time to death. You can use PROC PHREG to put gender into your model like this:

```
ods output ParameterEstimates = pvalue;

proc phreg;
    model daystodeath * deathcensor(0) = treatment gender;
run;

ods output close;
```

Note that with PROC PHREG all covariates need to be numeric, so "treatment" and "gender" need to be numeric. The *p*-values and hazard ratios that are useful for your statistical tables can be found in the "ProbChiSq" and "HazardRatio" variables, respectively, in the "pvalue" data set.

Obtaining Correlation Coefficients

On occasion you need to obtain *correlation coefficients* between two variables. Correlation coefficients are a way of measuring linear relationships between two variables. A correlation coefficient of 1 or –1 indicates a perfect linear relationship, and a coefficient of 0 indicates no strong linear relationship. Pearson correlation coefficients are useful for continuous variables, while Spearman correlation coefficients are useful for ordinal variables. For example, look at the following SAS code:

```
proc corr
    outp = pearson;
    var age weight;
run;

proc corr
    outs = spearman;
    var race treatment_success;
run;
```

The first PROC CORR sends the Pearson correlation coefficients to a data set called "pearson" for the continuous variables Age and Weight, while the second PROC CORR sends the Spearman correlation coefficients to a data set called "spearman" for the categorical variables Race and Treatment Success. The correlation coefficients are found where the "_TYPE_" variable is equal to "CORR" in the "pearson" and "spearman" data sets.

General Approach to Obtaining Statistics

The previous sections show you how to extract *p*-values for a commonly used set of statistical tests. This section describes a general step-by-step approach for getting your statistics from a SAS procedure into data sets for clinical trial table or graph reporting. Here are the steps to follow:

1. Determine what statistics you need in your table by looking at the listing destination output of your statistical procedure.

2. Check the SAS procedure syntax to see if there is an output data set that will provide you with the values you need. The output data sets from the SAS procedures are usually friendlier to use than the ODS OUTPUT data sets.

3. If you cannot find what you need in an output data set from the statistical procedure, use ODS OUTPUT to send your statistics to a data set. To determine

the name of the data set to output, perform an ODS TRACE on your SAS procedure like this:

```
ods trace on;
proc ...
run;
ods trace off;
```

Then go to your SAS log to see what "tables" the SAS procedure makes. Each block of text in your SAS listing output typically translates into a SAS data set in ODS. You can see what each table is called by looking at the "Output Added" blocks in your SAS log. These blocks look something like this:

```
Output Added:
-------------
Name:       ShortName
Label:      Dataset Label
Template:   3 level name
Path:       2 level name
-------------
```

4. "ShortName" from step 3 is what your ODS table name is called. Simply wrap an ODS OUTPUT statement around your SAS procedure to create the needed data set:

```
ods output ShortName = your_dataset_name;
proc ...
run;
ods output close;
```

The statistics you need are now in the data set called "your_dataset_name."

Note that when you obtain statistics from an ODS output data set, the results that you see there may appear different from what you see in your ODS listing destination (LST file). This is because a SAS procedure may round to a different precision in the ODS listing destination from the precision at which you present your ODS output statistics. The numbers are the same, but the way they are rounded may make the statistic appear different.

Chapter **8**

Exporting Data

Once your analyses and reporting are complete, you will probably need to export your SAS data to someone else. For clinical trials that are part of an FDA submission, the data export requirements are defined in the code of federal regulations and in FDA guidance. For clinical trials intended for other institutions, the data export requirements may vary widely. In this chapter we look at the various data export requirements and how you can use SAS to create data files for export.

Exporting Data to the FDA

Using the SAS XPORT Transport Format

The FDA guidance document "Providing Regulatory Submissions in Electronic Format –
General Considerations" as well as the eCTD guidance specifies that electronic data files
should be submitted to the FDA in SAS XPORT engine transport format. The SAS
XPORT transport format, historically known as the SAS 5 transport format, is an open
data standard published by SAS. (The open standard is documented fully in SAS
technical support document "TS-140: The Record Layout of a Data Set in SAS Transport
(XPORT) Format," found at http://support.sas.com/techsup/technote/ts140.html.) You
can create a SAS XPORT transport file by using the DATA step or PROC COPY with a
special libref. The following example shows both methods of export.

Program 8.1 Creating SAS XPORT Transport Format Data Sets for the FDA

```
libname sdtm "C:\sdtm_data";
libname dm xport "C:\dm.xpt";

**** PROC COPY METHOD TO CREATE A TRANSPORT FILE.;
proc copy
   in = sdtm
   out = dm;
      select dm;
run;

**** DATA STEP METHOD TO CREATE A TRANSPORT FILE.;
data dm.dm;
   set sdtm.dm;
run;
```

Both of these export methods create a SAS XPORT format transport file called dm.xpt.
Although any file extension could be used, "xpt" is the FDA-expected file extension for
the transport file. Also note that the libref "dm" is strange in that it contains an actual
filename and not just a directory path. This makes creating a series of SAS XPORT files
a bit more challenging. You could create a series of XPORT files with a SAS macro such
as the following.

Program 8.2 Creating Several SAS XPORT Transport Format Data Sets

```
***** THIS SAS MACRO CREATES A SERIES OF SAS XPORT FILES.;
***** PARAMETERS:  libname = raw data libref;
*****             dset    = name of data set;
%macro makexpt(libname=, dset=);

    libname &dset xport "c:\&dset..xpt";

    proc copy
       in = &libname
       out = &dset;
          select &dset;
    run;

%mend makexpt;

**** MAKEXPT CALLS;
%makexpt(libname = sdtm, dset = dm)
%makexpt(libname = sdtm, dset = ae)
...
```

SAS also provides a SAS macro that converts a directory of files from and to SAS XPORT format files. This SAS macro can be found at http://www.sas.com/govedu/fda/macro.html.

There are some additional concerns to be aware of when creating SAS XPORT format files. You should know that PROC CPORT can also create a SAS transport file, but the FDA does not support this proprietary format, so you should not use PROC CPORT for FDA data transfers. The LIBNAME statement in SAS lets you place more than one SAS data set into an XPORT format file, but for FDA submission you should place only one data set into each XPORT format file. Also, be careful when creating SAS data sets using SAS 7 or later versions. Because the SAS XPORT format is compliant with SAS 5 data sets, that means you are limited by SAS 5 data set constraints. Even though more current versions of SAS allow for longer labels, variable names, and variable lengths, you need to limit your SAS labels to 40 bytes, your variable names to 8 bytes, and your character variable widths to 200 bytes. If you exceed the SAS 5 data set constraints, your PROC COPY or DATA step code may terminate with errors or warnings when you attempt to create XPORT format files. Further details of the SAS XPORT format and FDA requirements can be found at http://www.sas.com/govedu/fda/index.html.

Creating XML Files

With the advent of the CDISC ODM model and the progression of the FDA's endorsement of the CDISC models, I believe that eventually all clinical trial data will likely be submitted to the FDA in ODM or a similar XML format. The XML-based ODM is already gaining acceptance within the pharmaceutical industry as a means of transferring clinical trial data. SAS provides two ways to produce ODM data files: using either PROC CDISC or the XML LIBNAME engine.

Using PROC CDISC

PROC CDISC is a SAS procedure available as a hot fix for SAS 8.2 that ships as part of SAS 9.1.3. PROC CDISC allows you to export (and import) XML files that are compliant with the CDISC ODM version 1.2 schema. Here are a sample PROC CONTENTS and a sample PROC PRINT of a SDTM data set called DM (for demographics) that we will export to ODM using PROC CDISC.

PROC CONTENTS of DM data set:

#	Variable	Type	Len	Label
	Alphabetic List of Variables and Attributes			
9	age	Num	8	Age in AGEU at Reference Date/Time
10	ageu	Char	6	Age Units
13	arm	Char	16	Description of Planned Arm
14	armcd	Num	8	Planned Arm Code
7	brthdtc	Char	16	Date/Time of Birth
12	country	Char	13	Country
8	dmdtc	Char	16	Date/Time of Collection
11	dmdy	Num	8	Study Day of Collection
5	domain	Char	2	Domain Abbreviation
17	race	Char	9	Race
16	rfendtc	Char	16	Subject Reference End Date/Time
6	rfstdtc	Char	16	Subject Reference Start Date/Time
15	sex	Char	1	Sex
3	siteid	Char	10	Study Site Identifier
1	studyid	Char	20	Study Identifier

(continued)

Alphabetic List of Variables and Attributes				
#	**Variable**	**Type**	**Len**	**Label**
4	subjid	Char	25	Subject Identifier for the Study
2	usubjid	Char	40	Unique Subject Identifier

PROC PRINT of DM data set:

Obs	studyid	usubjid	siteid	subjid	domain	rfstdtc	brthdtc	dmdtc	age	ageu	dmdy	country	arm	armcd	sex	rfendtc	race
1	XT802	802101001	101	802101001	DM	2003-01-10T12:52	1975-02-26	2003-01-10	27	YEARS	1	United States	Drug B	1	M	2003-04-12T10:00	CAUCASIAN
2	XT802	802101002	101	802101002	DM	2003-01-10T18:40	1950-01-30	2003-01-10	52	YEARS	1	United States	Drug A	0	F	2003-04-11T12:33	CAUCASIAN
3	XT802	802101003	101	802101003	DM	2003-02-24T17:00	1944-02-21	2003-02-24	59	YEARS	1	United States	Drug B	1	M	2003-05-22T13:25	CAUCASIAN
4	XT802	802101004	101	802101004	DM	2003-02-24T11:01	1967-03-16	2003-02-24	35	YEARS	1	United States	Drug A	0	M	2003-05-12T11:33	CAUCASIAN
5	XT802	802101005	101	802101005	DM	2003-02-24T12:51	1964-01-22	2003-02-24	38	YEARS	1	United States	Drug B	1	M	2003-06-04T14:47	CAUCASIAN
6	XT802	802101006	101	802101006	DM	2003-02-25T10:30	1977-11-16	2003-02-25	25	YEARS	1	United States	Drug A	0	M	2003-05-25T15:33	CAUCASIAN
7	XT802	802101007	101	802101007	DM	2003-03-01T19:42	1969-06-22	2003-03-01	33	YEARS	1	United States	Drug B	1	F	2003-06-04T11:47	CAUCASIAN
8	XT802	802101008	101	802101008	DM	2003-03-02T12:33	1949-05-14	2003-03-02	53	YEARS	1	United States	Drug A	0	M	2003-06-12T12:52	CAUCASIAN
9	XT802	802101009	101	802101009	DM	2003-03-03T19:13	1971-02-24	2003-03-03	31	YEARS	1	United States	Drug B	1	M	2003-04-02T21:12	CAUCASIAN
10	XT802	802101010	101	802101010	DM	2003-03-03T12:43	1955-04-29	2003-03-03	47	YEARS	1	United States	Drug A	0	F	2003-05-12T20:42	CAUCASIAN

To export this file to ODM using PROC CDISC, you run the following SAS program. It is annotated for further discussion.

Program 8.3 Using PROC CDISC to Create an ODM XML File

```
libname sdtm "C:\sdtm_data";

**** SPECIFY PROC CDISC PARAMETERS IN DATA STEPS.;          ❶
data odm;
   ODMVersion = "1.2";
   fileOID = "2004-09-11 Transfer of XT802";
   FileType = "Snapshot";
run;

data study;
   StudyOID = "XT802";
run;

data globalvariables;
   StudyName = "XT802";
   StudyDescription =
      "Clinical Trial XT802 - Study of Infectious Agent";
   ProtocolName = "INVAG-XT802";
run;

data metadataversion;
   MetadataVersionOID = "SDTMV3.1_01.00";
   Name = "Submissions Data Tabulation Model Version 3.1 -
Trial Meta Version 01.00";
run;

**** GET DM DATA SET AND DEFINE 10 KEYSET VARIABLES.;
data dm;                                                    ❷
   length __STUDYOID __METADATAVERSIONOID __STUDYEVENTOID
          __SUBJECTKEY __FORMOID __ITEMGROUPOID
          __ITEMGROUPREPEATKEY __TRANSACTIONTYPE
          __STUDYEVENTREPEATKEY __FORMREPEATKEY $ 100.;

   set sdtm.dm;

   retain __STUDYOID              "TRIALXT802"
          __METADATAVERSIONOID "SDTMV3.1_01.00"
          __STUDYEVENTOID         "BASELINE"
          __FORMOID               "DM"
          __ITEMGROUPOID          "DM"
          __ITEMGROUPREPEATKEY "1"
          __TRANSACTIONTYPE       "Snapshot"
          __STUDYEVENTREPEATKEY "1"
          __FORMREPEATKEY         "1";
```

```
    **** MAP __SUBJECTKEY NEEDED FOR EXPORT TO USUBJID.;
    __SUBJECTKEY = usubjid;
run;

filename xmlout "C:\ODM_FILES\dm.xml";

**** EXPORT ODM DM FILE.;
proc cdisc
   model = odm
   write = xmlout;

   odm                data = work.odm;
   study              data = work.study;
   globalvariables    data = work.globalvariables;
   metadataversion    data = work.metadataversion;
   clinicaldata       data = work.DM
                       domain  = "DM"
                       name    = "Demographics"
                       comment = "Patient Demographics";
run;
```

Notes for Program 8.3:

❶ PROC CDISC requires a number of parameters that specify the clinical trial
    metadata not typically found in your SAS data sets. These parameters can be
    specified within PROC CDISC or in separate data sets that can be passed to the
    procedure. This example chooses the latter method.

❷ You need to define the key set variables for PROC CDISC to export your data to
    XML. ODM specifications have a maximum length of 100 characters for these
    fields, so here I have set the length of the key set fields to 100. It is important to
    understand what these key set fields are in order to set them properly. In brief, a
    STUDYEVENT is essentially a visit, and many FORMs can be attributed to a
    STUDYEVENT. A FORM is equivalent to a CRF page. An ITEMGROUP is a
    group of variables that make up a discrete piece or all of a FORM. The
    REPEATKEY fields indicate whether there are multiple observations within a
    STUDYEVENT, FORM, or ITEMGROUP. For the DM file all of the
    REPEATKEY fields are set to "1." Note how __SUBJECTKEY is defined by the
    USUBJID variable.

The subsequent slightly truncated dm.xml file looks like this:

```
<?xml version="1.0" encoding="windows-1252" ?>
<!--
      Clinical Data Interchange Standards Consortium (CDISC)
      Operational Data Model (ODM) for clinical data interchange

      You can learn more about CDISC standards efforts at
      http://www.cdisc.org/standards/index.html
  -->

<ODM xmlns="http://www.cdisc.org/ns/odm/v1.2"
     xmlns:ds="http://www.w3.org/2000/09/xmldsig#"
     xmlns:xsi="http://www.w3.org/2001/XMLSchema-instance"
     xsi:schemaLocation="http://www.cdisc.org/ns/odm/v1.2 ODM1-2-0.xsd"

     ODMVersion="1.2"
     FileOID="2004-09-11 Transfer of XT802"
     FileType="Snapshot"

     AsOfDateTime="2004-10-02T09:06:11"
     CreationDateTime="2004-10-02T09:06:11"
     SourceSystem="SAS 9.1"
     SourceSystemVersion="9.01.01M3P07282004">

   <Study OID="XT802">

     <!--
          GlobalVariables is a REQUIRED section in ODM markup
       -->
     <GlobalVariables>
        <StudyName>XT802</StudyName>
        <StudyDescription>Clinical Trial XT802 - Study of Infectious Agent
                    </StudyDescription>
        <ProtocolName>INVAG-XT802</ProtocolName>
     </GlobalVariables>

     <BasicDefinitions />

     <!--
          Internal ODM markup required metadata
       -->
     <MetaDataVersion OID="SDTMV3.1_01.00"
     Name="Submissions Data Tabulation Model Version 3.1 - Trial Meta
          Version 01.00">
        <Protocol>
```

```
      <StudyEventRef StudyEventOID="BASELINE" OrderNumber="1"
                                  Mandatory="Yes" />
</Protocol>

<StudyEventDef OID="BASELINE" Name="Study Event Definition"
                                  Repeating="No"
 Type="Common">
    <FormRef FormOID="DM" OrderNumber="1" Mandatory="No" />
</StudyEventDef>

<FormDef OID="DM" Name="Form Definition" Repeating="No">
    <ItemGroupRef ItemGroupOID="DM" Mandatory="No" />
</FormDef>


<!--
      Columns defined in the table
  -->
<ItemGroupDef OID="DM" Repeating="No"
              SASDatasetName="DM"
              Name="Demographics"
              Domain="DM"
              Comment="Patient Demographics">
    <ItemRef ItemOID="ID.studyid" OrderNumber="1" Mandatory="No" />
    <ItemRef ItemOID="ID.usubjid" OrderNumber="2" Mandatory="No" />
    <ItemRef ItemOID="ID.siteid" OrderNumber="3" Mandatory="No" />
    <ItemRef ItemOID="ID.subjid" OrderNumber="4" Mandatory="No" />
    <ItemRef ItemOID="ID.domain" OrderNumber="5" Mandatory="No" />
    <ItemRef ItemOID="ID.rfstdtc" OrderNumber="6" Mandatory="No" />
    <ItemRef ItemOID="ID.brthdtc" OrderNumber="7" Mandatory="No" />
    <ItemRef ItemOID="ID.dmdtc" OrderNumber="8" Mandatory="No" />
    <ItemRef ItemOID="ID.age" OrderNumber="9" Mandatory="No" />
    <ItemRef ItemOID="ID.ageu" OrderNumber="10" Mandatory="No" />
    <ItemRef ItemOID="ID.dmdy" OrderNumber="11" Mandatory="No" />
    <ItemRef ItemOID="ID.country" OrderNumber="12" Mandatory="No" />
    <ItemRef ItemOID="ID.arm" OrderNumber="13" Mandatory="No" />
    <ItemRef ItemOID="ID.armcd" OrderNumber="14" Mandatory="No" />
    <ItemRef ItemOID="ID.sex" OrderNumber="15" Mandatory="No" />
    <ItemRef ItemOID="ID.rfendtc" OrderNumber="16" Mandatory="No" />
    <ItemRef ItemOID="ID.race" OrderNumber="17" Mandatory="No" />
</ItemGroupDef>
```

```
<!--
     Column attributes as defined in the table
  -->
<ItemDef OID="ID.studyid" SASFieldName="studyid" Name="Study
              Identifier"
 DataType="text" Length="20" />
<ItemDef OID="ID.usubjid" SASFieldName="usubjid"
 Name="Unique Subject Identifier" DataType="text" Length="40" />
<ItemDef OID="ID.siteid" SASFieldName="siteid" Name="Study Site
              Identifier"
 DataType="text" Length="10" />
<ItemDef OID="ID.subjid" SASFieldName="subjid"
 Name="Subject Identifier for the Study" DataType="text" Length="25"
      />
<ItemDef OID="ID.domain" SASFieldName="domain" Name="Domain
              Abbreviation"
 DataType="text" Length="2" />
<ItemDef OID="ID.rfstdtc" SASFieldName="rfstdtc"
 Name="Subject Reference Start Date/Time" DataType="text"
      Length="16" />
<ItemDef OID="ID.brthdtc" SASFieldName="brthdtc" Name="Date/Time of
              Birth"
 DataType="text" Length="16" />
<ItemDef OID="ID.dmdtc" SASFieldName="dmdtc" Name="Date/Time of
              Collection"
 DataType="text" Length="16" />
<ItemDef OID="ID.age" SASFieldName="age"
 Name="Age in AGEU at Reference Date/Time" DataType="float" />
<ItemDef OID="ID.ageu" SASFieldName="ageu" Name="Age Units"
 DataType="text" Length="6" />
<ItemDef OID="ID.dmdy" SASFieldName="dmdy" Name="Study Day of
              Collection"
 DataType="float" />
<ItemDef OID="ID.country" SASFieldName="country" Name="Country"
 DataType="text" Length="13" />
<ItemDef OID="ID.arm" SASFieldName="arm" Name="Description of
              Planned Arm"
 DataType="text" Length="16" />
<ItemDef OID="ID.armcd" SASFieldName="armcd" Name="Planned Arm Code"
 DataType="float" />
<ItemDef OID="ID.sex" SASFieldName="sex" Name="Sex" DataType="text"
 Length="1" />
<ItemDef OID="ID.rfendtc" SASFieldName="rfendtc"
 Name="Subject Reference End Date/Time" DataType="text" Length="16"
      />
<ItemDef OID="ID.race" SASFieldName="race" Name="Race"
              DataType="text"
```

```
        Length="9" />
    </MetaDataVersion>
</Study>


<!--
     Administrative metadata
  -->
<AdminData />


<!--
     Clinical Data    : DM
                        Demographics
                        Patient Demographics
  -->
<ClinicalData StudyOID="TRIALXT802" MetaDataVersionOID="SDTMV3.1_01.00">
    <SubjectData SubjectKey="802101001">
        <StudyEventData StudyEventOID="BASELINE" StudyEventRepeatKey="1">
            <FormData FormOID="DM" FormRepeatKey="1">
                <ItemGroupData ItemGroupOID="DM" ItemGroupRepeatKey="1"
                  TransactionType="Snapshot">
                    <ItemData ItemOID="ID.studyid" Value="XT802" />
                    <ItemData ItemOID="ID.usubjid" Value="802101001" />
                    <ItemData ItemOID="ID.siteid" Value="101" />
                    <ItemData ItemOID="ID.subjid" Value="802101001" />
                    <ItemData ItemOID="ID.domain" Value="DM" />
                    <ItemData ItemOID="ID.rfstdtc" Value="2003-01-10T12:52" />
                    <ItemData ItemOID="ID.brthdtc" Value="1975-02-26" />
                    <ItemData ItemOID="ID.dmdtc" Value="2003-01-10" />
                    <ItemData ItemOID="ID.age" Value="27" />
                    <ItemData ItemOID="ID.ageu" Value="YEARS" />
                    <ItemData ItemOID="ID.dmdy" Value="1" />
                    <ItemData ItemOID="ID.country" Value="United States" />
                    <ItemData ItemOID="ID.arm" Value="Drug B" />
                    <ItemData ItemOID="ID.armcd" Value="1" />
                    <ItemData ItemOID="ID.sex" Value="M" />
                    <ItemData ItemOID="ID.rfendtc" Value="2003-04-12T10:00" />
                    <ItemData ItemOID="ID.race" Value="CAUCASIAN" />
                </ItemGroupData>
            </FormData>
        </StudyEventData>
    </SubjectData>
    <SubjectData SubjectKey="802101002">
        <StudyEventData StudyEventOID="BASELINE" StudyEventRepeatKey="1">
            <FormData FormOID="DM" FormRepeatKey="1">
                <ItemGroupData ItemGroupOID="DM" ItemGroupRepeatKey="1"
```

```
            TransactionType="Snapshot">
              <ItemData ItemOID="ID.studyid" Value="XT802" />
              <ItemData ItemOID="ID.usubjid" Value="802101002" />
              <ItemData ItemOID="ID.siteid" Value="101" />
              <ItemData ItemOID="ID.subjid" Value="802101002" />
              <ItemData ItemOID="ID.domain" Value="DM" />
              <ItemData ItemOID="ID.rfstdtc" Value="2003-01-10T18:40" />
              <ItemData ItemOID="ID.brthdtc" Value="1950-01-30" />
              <ItemData ItemOID="ID.dmdtc" Value="2003-01-10" />
              <ItemData ItemOID="ID.age" Value="52" />
              <ItemData ItemOID="ID.ageu" Value="YEARS" />
              <ItemData ItemOID="ID.dmdy" Value="1" />
              <ItemData ItemOID="ID.country" Value="United States" />
              <ItemData ItemOID="ID.arm" Value="Drug A" />
              <ItemData ItemOID="ID.armcd" Value="0" />
              <ItemData ItemOID="ID.sex" Value="F" />
              <ItemData ItemOID="ID.rfendtc" Value="2003-04-11T12:33" />
              <ItemData ItemOID="ID.race" Value="CAUCASIAN" />
            </ItemGroupData>
          </FormData>
        </StudyEventData>
      </SubjectData>
      .
      .
      .
   ...The remaining 8 observations of the DM data set would repeat here
      .
      .
      .

      </ClinicalData>
  </ODM>
```

Complete SAS documentation for PROC CDISC can be found at
http://support.sas.com/rnd/base/topics/sxle82/TW8774.pdf. In the future, it is expected
that PROC CDISC will be able to export multiple SAS data sets to a single ODM XML
file. As the CDISC models and PROC CDISC rapidly evolve, continue to watch PROC
CDISC as a tool for CDISC model conversion.

Using the XML LIBNAME Engine

The other option for producing ODM XML is to use the XML LIBNAME engine. You
can convert the previous DM SAS data set to ODM version 1.2 by running the following
SAS program.

Program 8.4 Using the XML LIBNAME Engine to Create an ODM XML File

```
libname   sdtm    'C:\sdtm_data';
filename output 'C:\dm.xml';
libname   output xml xmltype = CDISCODM FormatActive = yes;


**** GET DM DATA SET AND DEFINE 10 KEYSET VARIABLES.;
data dm;
    length __STUDYOID __METADATAVERSIONOID __STUDYEVENTOID
           __SUBJECTKEY __FORMOID __ITEMGROUPOID
           __ITEMGROUPREPEATKEY __TRANSACTIONTYPE
           __STUDYEVENTREPEATKEY __FORMREPEATKEY $ 100.;

    set sdtm.dm;

    retain __STUDYOID             "TRIALXT802"
           __METADATAVERSIONOID "SDTMV3.1_01.00"
           __STUDYEVENTOID      "BASELINE"
           __FORMOID            "DM"
           __ITEMGROUPOID       "DM"
           __ITEMGROUPREPEATKEY "1"
           __TRANSACTIONTYPE    "Snapshot"
           __STUDYEVENTREPEATKEY "1"
           __FORMREPEATKEY       "1";

   **** MAP __SUBJECTKEY NEEDED FOR EXPORT TO USUBJID;
   __SUBJECTKEY = usubjid;
run;

**** EXPORT DM ODM FILE.;
data output.dm;
    set dm;
run;
```

The resulting ODM XML file produced by the XML LIBNAME engine is very similar to the one produced by PROC CDISC. However, PROC CDISC allows you to specify and pass more of the metadata information along to the XML file. If you need to create customized XML files for a sponsor that do not match the ODM specification, you can use the SAS XML Mapper and the XML LIBNAME engine to write your own custom XML files.

Exporting Data Not Destined for the FDA

Clinical trial data that is not sent to the FDA can be exported from SAS in many ways other than SAS XPORT format files or ODM XML. This section discusses the numerous ways to export data from SAS to entities other than the FDA.

Exporting Data with PROC CPORT

An excellent way to export information from SAS to another SAS user is with PROC CPORT. PROC CPORT creates an operating system–independent SAS transport file that can hold data sets as well as SAS programs, permanent format catalogs, and other SAS catalogs. Note that PROC CPORT files are not backward compatible with older versions of SAS, so a transport file created by PROC CPORT in SAS 9 cannot be read by SAS 8. Here is some example code showing how to place all of the SAS data sets and permanent formats from a libref called "library" into a SAS transport file called "mytrial.xpt."

Program 8.5 Using PROC CPORT to Create a SAS Transport File

```
libname library "c:\mytrial";
filename tranfile 'c:\mytrial.xpt';

**** COPY ALL SAS DATA SETS AND PERMANENT FORMATS FROM LIBRARY
**** INTO MYTRIAL.XPT.;
proc cport
   library = library
   file = tranfile
   exclude sasmacr;
run;
```

Exporting ASCII Text

On occasion you will need to export your SAS data as ASCII text. You may find that the recipient of your data requires that you send some form of delimited ASCII text, because that is all their software can read. In fact, delimited ASCII text is a primitive data exchange format that almost any software can read.

Using the Export Wizard and PROC EXPORT

PROC EXPORT provides a quick way to create ASCII text files from SAS data sets. You can call PROC EXPORT by typing in the SAS code, or you can use the convenient SAS windowing environment Export Wizard that builds the PROC EXPORT code for you. Let's start by looking at using the SAS 9.1 windowing environment Export Wizard to export the DM demographics data set used earlier as a comma-separated values (CSV) text file. You begin the process in the SAS windowing environment by selecting "File" from the toolbar and then "Export Data" from the menu. You will see a Select Library and Member window that looks like this:

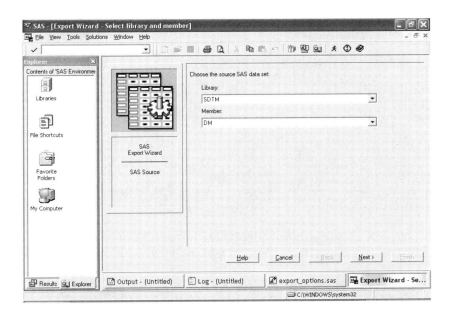

Select the "Library" (libref) and "Member" (data set) to be exported, and then click "Next" to open the Select Export Type window. To get a CSV file, select "Comma Separated Values" as the standard data source, as follows:

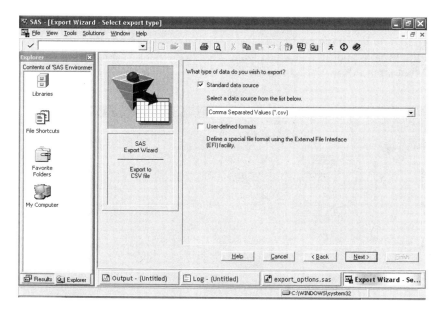

When you click "Next," a Select File window opens for you to choose the file destination. Click "Browse" to browse to the destination file, or enter the path to the file as follows:

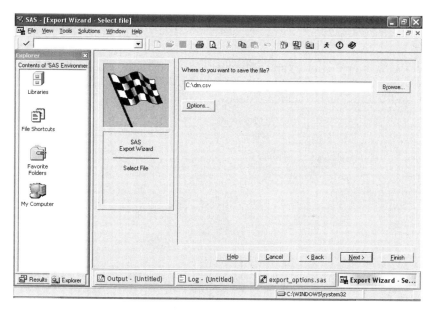

Because we explicitly chose comma-separated values, the "Options" button does nothing here. We will revisit that in a moment. Click "Next" to open the Create SAS Statements window, where you can save the PROC EXPORT code:

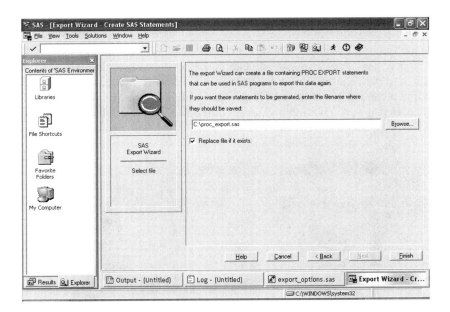

If you click "Finish," the resulting dm.csv CSV file looks like this:

```
studyid,usubjid,siteid,subjid,domain,rfstdtc        ...
XT802,802101001,101,802101001,DM,2003-01-10T12:52   ...
XT802,802101002,101,802101002,DM,2003-01-10T18:40   ...
XT802,802101003,101,802101003,DM,2003-02-24T17:00   ...
XT802,802101004,101,802101004,DM,2003-02-24T11:01   ...
XT802,802101005,101,802101005,DM,2003-02-24T12:51   ...
XT802,802101006,101,802101006,DM,2003-02-25T10:30   ...
XT802,802101007,101,802101007,DM,2003-03-01T19:42   ...
XT802,802101008,101,802101008,DM,2003-03-02T12:33   ...
XT802,802101009,101,802101009,DM,2003-03-03T19:13   ...
XT802,802101010,101,802101010,DM,2003-03-03T12:43   ...
```

Here is the PROC EXPORT code in "proc_export.sas" created by the Export Wizard:

```
PROC EXPORT DATA= SDTM.DM
            OUTFILE= "C:\dm.csv"
            DBMS=CSV REPLACE;
RUN;
```

There are other options for creating delimited ASCII files from the Export Wizard. From the Select Export Type window you can pick "tab delimited" or "delimited file" as one of the standard data sources if you want delimited files. If you choose the "delimited file" option, you can choose your field delimiter by clicking the "Options" button. For instance, if you were creating a pipe-delimited file, the Select Export Type window would look like this:

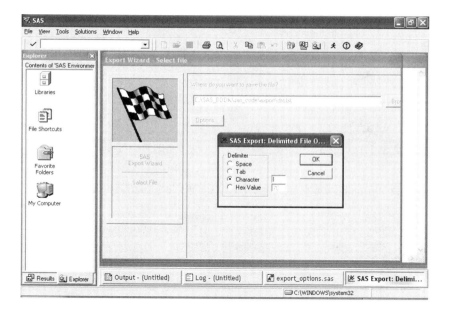

You can see how the "|" (pipe) character is entered as the delimiter in the "Options" dialog box. If you have more complex requirements for the ASCII text file you want to export, you can invoke "External File Interface" in the Select Export Type window, write customized DATA step code with FILE and PUT statements, or use some of the ODS tagsets supplied by SAS, found at http://support.sas.com/rnd/base/topics/odsmarkup/, that have the ability to create numerous types of ASCII text formats.

Using SAS Enterprise Guide to Export Data

Another approach to exporting text files with SAS is to use the SAS Enterprise Guide interface. The following example exports the DM data set to a CSV file using SAS Enterprise Guide 3.0. With the DM data set selected in the Process Flow window, click "File" and then select "Export dm" from the drop-down list as follows:

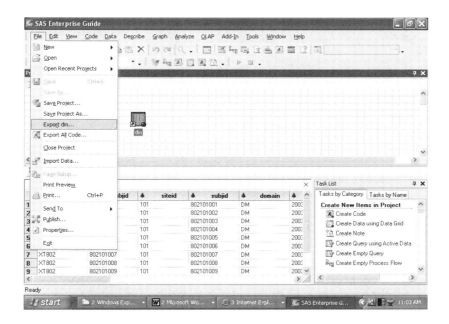

A pop-up dialog box will ask where you want to save the file. Select "Local Computer" or "SAS Server," as desired. Then a pop-up "Export" dialog box appears, where you select "CSV (Comma delimited)" as follows:

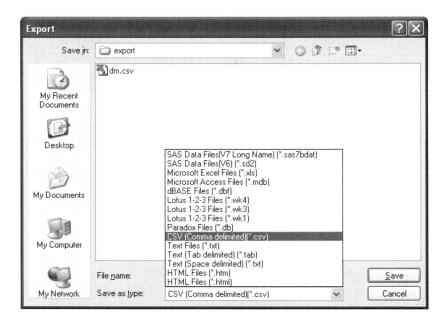

Here is the resulting dm.csv CSV file created by SAS Enterprise Guide:

```
"studyid","usubjid","siteid","subjid","domain","rfstdtc" ...
"XT802","802101001","101","802101001","DM","2003-01-10T12:52" ...
"XT802","802101002","101","802101002","DM","2003-01-10T18:40" ...
"XT802","802101003","101","802101003","DM","2003-02-24T17:00" ...
"XT802","802101004","101","802101004","DM","2003-02-24T11:01" ...
"XT802","802101005","101","802101005","DM","2003-02-24T12:51" ...
"XT802","802101006","101","802101006","DM","2003-02-25T10:30" ...
"XT802","802101007","101","802101007","DM","2003-03-01T19:42" ...
"XT802","802101008","101","802101008","DM","2003-03-02T12:33" ...
"XT802","802101009","101","802101009","DM","2003-03-03T19:13" ...
"XT802","802101010","101","802101010","DM","2003-03-03T12:43" ...
```

Note that the CSV file generated by SAS Enterprise Guide 3.0 is slightly different from the CSV file generated by PROC EXPORT. SAS Enterprise Guide wraps the variable contents in double quotation marks regardless of whether a comma appears in the data. SAS Enterprise Guide can also easily export your SAS data to tab- or space-delimited ASCII text files. SAS Enterprise Guide does not have the External File Interface found in the SAS windowing environment, which you may need for more complex ASCII file exports.

Exporting Data to Microsoft Office Files

Because the Microsoft Office suite is so widely used, it is sometimes necessary for you to export SAS data sets into Microsoft Access or Microsoft Excel files. SAS provides the Export Wizard/PROC EXPORT and the SAS Enterprise Guide interface for exporting data directly to Microsoft Office files.

Using the Export Wizard and PROC EXPORT

The SAS windowing environment Export Wizard provides an easy way to export the contents of a SAS data set to a Microsoft Excel or Access file. Here again, the Export Wizard is a graphical user interface that builds the PROC EXPORT code for you. Let's look at an example of how to export the DM demographics data directly to an Excel file. You begin in the SAS windowing environment by selecting "File" from the toolbar and then "Export Data" from the menu. The Select Library and Member window pops up, allowing you to pick the library and SAS data set to export.

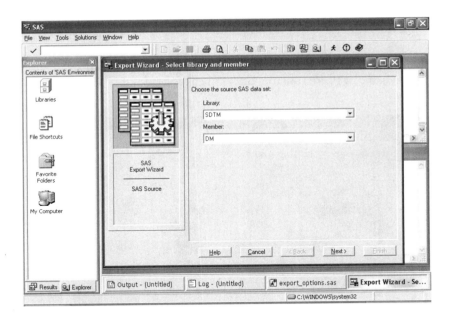

After you pick your data set, click "Next," and a file browser window pops up that allows you to select the Microsoft Excel version you want. Here we select an Excel 2002 file.

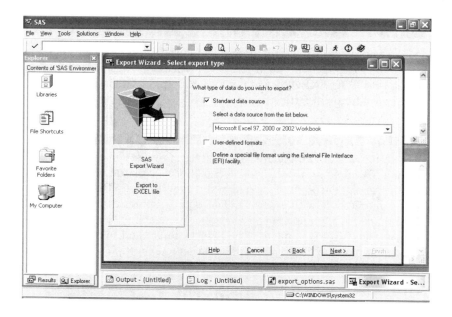

After you click "Next," the "Connect to MS Excel" dialog box appears, allowing you to specify the name of the destination Excel file as follows:

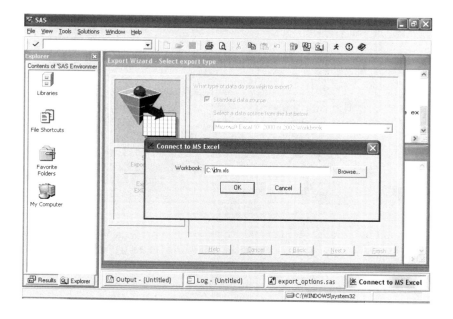

After you click "Next," a "Select Table" window pops up, allowing you to enter a name for your Excel worksheet. We call the worksheet "DM" in this example, as follows:

Click "Next," and SAS offers to save the PROC EXPORT statements in a SAS program. Click "Finish" and your Excel file is created. The PROC EXPORT code SAS saved in this example follows:

```
PROC EXPORT DATA= SDTM.DM
            OUTFILE= "C:\dm.xls"
            DBMS=EXCEL REPLACE;
      SHEET="DM";
RUN;
```

The Export Wizard process for Microsoft Access files works like the one for Excel files and produces similar PROC EXPORT code. The previous DM SAS data set exported to Microsoft Access with the Export Wizard creates this PROC EXPORT code:

```
PROC EXPORT DATA= SDTM.DM
            OUTTABLE= "dm"
            DBMS=ACCESS REPLACE;
      DATABASE="C:\dm.mdb";
RUN;
```

Using SAS Enterprise Guide to Export Data

SAS Enterprise Guide 3.0 offers a simple, convenient export tool for Microsoft Access and Excel files. To export the DM data set to Microsoft Excel in SAS Enterprise Guide, simply select the file in the Process Flow window, click "File," and select "Export dm" from the menu as follows:

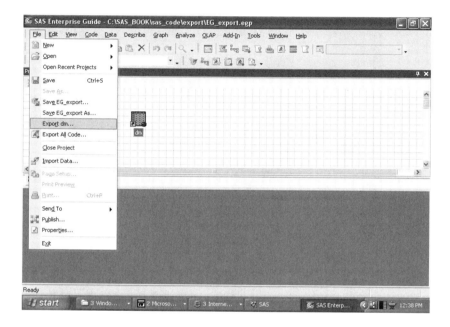

Select "Local Computer" or "SAS Servers" from the pop-up dialog box. Then, select "Microsoft Excel Files" from the "Save as type" selection list in the Export window as follows:

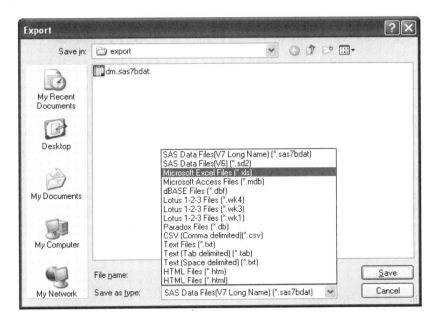

Your dm.xls Excel file is then created. The process of exporting a Microsoft Access database file is just as easy, requiring only a few clicks of the mouse.

Exporting Other Proprietary Data Formats

You may find that you need to export your SAS data as something other than regular ASCII text or Microsoft Office files. In this case, the export wizards in the SAS windowing environment in SAS 9 and SAS Enterprise Guide 3.0 can easily export the following file formats:

Database	SAS 9 DMS	SAS Enterprise Guide
dBASE	Yes	Yes
Lotus 1-2-3	Yes	Yes
Paradox	Yes	Yes
HTML	No	Yes
JMP	Yes	No

If you have a need to convert SAS data sets to other file formats that SAS cannot handle directly, you might consider using a product such as DBMS/Copy. DBMS/Copy is a software package distributed by the SAS subsidiary Dataflux. See the end of Chapter 3 for the list of databases that DBMS/Copy can create from your SAS data sets.

Encryption and File Transport Options

Once you have your SAS data ready for transport, you need to determine a means to deliver it. There are many ways to send data, but you should strive for process simplicity and data security. To keep your data secure and to comply with 21 CFR–Part 11, you need to encrypt your data files for transport. The best encryption you can use is key exchange high-bit encryption software such as *PGP*, which creates essentially unbreakable files when used properly. Once your data files are encrypted, you can either send them on physical media such as CD-ROM or send them electronically with secure transmission software such as Secure File Transport Protocol (SFTP). If you need to send data to someone once, a CD-ROM is simple enough to produce. However, if you need to send the data repeatedly, then you should use a more automated electronic method of data exchange. Shell scripts and batch files can be written to automate the electronic data transfer process.

C h a p t e r 9

The Future of SAS Programming in Clinical Trials

This chapter briefly discusses the future of SAS programming and the SAS programmer in the world of clinical trials. The impact of the business environment, regulatory issues, technological changes, and standards organizations on the future of SAS programming in clinical trials is addressed.

Changes in the Business Environment

Like many industries, the pharmaceutical industry is under strong financial pressure to increase business efficiency. The competitive forces of a capitalist marketplace create pressure to be more efficient. Your company needs to outperform the competition in order to survive. Further pressure comes from the political and social perception that medicines and medical devices are too expensive. Pharmaceutical companies are constantly working to bring their costs down in order to lessen the wrath of the drug-consuming public. Finally, the global economy is having an effect on the pharmaceutical industry, and large pharmaceutical companies and contract research organizations are moving their operations to where labor is cheaper. This changing business environment in the pharmaceutical industry means that, as a SAS programmer, you need to bring your SAS programming and technical skills to bear to help your company be as efficient as possible.

Changes in Technology

In some ways, the technological environment has changed dramatically in the clinical trial business. In 1989 Tim Berners-Lee invented the World Wide Web. Since then there has been explosive development of Internet and electronic communications technology. The clinical trial business has slowly been integrating this new wave of information technology into its processes. In the past, regulatory submissions to the authorities were entirely paper based, and a New Drug Application could fill a small truck. Currently drug applications are being sent to the authorities partially or entirely in electronic format, consisting of paper documents, PDF documents, and SAS transport format files. In the future the drug application and approval process will likely be completely electronic. Perhaps your entire drug application will be sent electronically to your regulatory authority as a standardized XML data stream. As a SAS programmer, you need to keep abreast of technological change so that you can continue to be a productive information technology professional.

Changes in Regulations

The regulatory environment for clinical trial research has changed over time. In 1997 the regulation for electronic signatures and electronic records, 21 CFR–Part 11, went into effect. This regulation has had a broad impact on the clinical trial industry, because it requires system validation and security of all clinical data software used in the conduct of clinical trials. In 2003 new HIPAA regulations from the United States Department of Health and Human Services placed privacy restrictions on how patient data can be shared. If you work for an entity covered by this regulation, the HIPAA privacy rule changes what clinical trial data you can share and with whom you can share it, and the HIPAA security rule places additional restrictions on data security that went into effect in 2005. Adding to these constraints are new regulations such as the eCTD that provide the opportunity for more electronic submissions to regulatory authorities. With the constantly changing regulatory environment, you as a SAS programmer need to stay abreast of the regulations so that you can keep your work and organization compliant.

Changes in Standards

Outside of regulations and regulatory guidance, the clinical trial industry has not adopted many standards. In the past, each pharmaceutical company or contract research organization may have had its own internal data standards and processes, but there were not many global data standards for clinical trials. This lack of standards led to inefficiency as companies had to negotiate how to exchange their clinical trial information. CDISC has begun to solve that problem by publishing open data models for the worldwide clinical trial industry to use. As the adoption of the CDISC data models grows, you will see software systems grow in their ability to use them. This will mean more standardization of data exchange tools and clinical reporting tools. As a SAS programmer, you need to be working toward standardizing your data analysis processes around the use of the data standards provided by CDISC.

292 SAS Programming in the Pharmaceutical Industry

Use of SAS Software in the Clinical Trial Industry

SAS has always had and will maintain a central role in the data management, analysis, and reporting of clinical trial data. Because of the strong suite of SAS statistical procedures and the power of Base SAS programming, SAS remains a favorite of statisticians for the analysis of clinical trial data. Several companies have built their clinical trial data management and statistical analysis systems entirely with SAS software. More recently, SAS has offered SAS Drug Development as an industry solution that provides a comprehensive clinical trial analysis and reporting environment compliant with 21 CRF–Part 11.

In the future, whether you use SAS Drug Development, SAS Enterprise Guide, the SAS windowing environment, or some other front end to SAS software, the raw power of SAS for clinical trial work will continue to derive from Base SAS, SAS/STAT, SAS/GRAPH, and the SAS macro language. Base SAS and the DATA step are exceedingly powerful tools for clinical data manipulation. SQL, commonly used with most relational databases, cannot readily handle some of the DATA step code required for complex clinical trial data manipulation and derivation. Because of the strength of these core SAS products and the continuing commitment of SAS to supporting the pharmaceutical industry, I believe that SAS will remain the premiere analysis tool for clinical trial analysis and reporting.

Chapter **10**

Further Resources

This book cannot cover the entire sum of knowledge that you will need to be a SAS programmer in the pharmaceutical industry, so this chapter provides some further resources for you to explore.

Regulatory Resources

The following resources will help you to write regulatory-compliant SAS applications.

SAS Programming Validation

Because SAS programming in the pharmaceutical industry is governed by 21 CRF–Part 11, your programming must be validated. This book does not speak much to the process of SAS programming validation primarily because that topic would fill a book by itself. (For a comprehensive look at the validation of SAS programming in the pharmaceutical industry, look for the forthcoming SAS Press book from Shilling and Matthews, tentatively titled *Validating Clinical Trials Using SAS.*) Here are some other resources for validation of SAS programming:

- "21 CFR–Part 11 Electronic Records; Electronic Signatures; Final Rule," found at http://www.fda.gov/ora/compliance_ref/part11/FRs/background/pt11finr.pdf.

- "General Principles of Software Validation; Final Guidance for Industry and FDA Staff," found at http://www.fda.gov/cdrh/comp/guidance/938.pdf.

- "Guide to Inspection of Computerized Systems in Drug Processing," found at http://www.fda.gov/ora/inspect_ref/igs/csd.html.

- "Computerized Systems Used in Clinical Trials," found at http://www.fda.gov/ora/compliance_ref/bimo/ffinalcct.pdf.

- The Association for Clinical Data Management (ACDM) and Statisticians in the Pharmaceutical Industry (PSI) publish an excellent document called "Computer Systems Validation in Clinical Research: A Practical Guide," which can be found at http://www.cr-csv.org/.

- The Web site at http://21cfrpart11.com/ is a central repository for issues regarding 21 CRF – Part 11.

- Dr. Joshua Sharlin and other consultants who can be found on the Internet offer classes and materials on validating SAS programming in the pharmaceutical industry. Dr. Sharlin's services can be found online at http://www.speedupfda.com.

FDA Resources

Besides 21 CFR – Part 11, the FDA has many other documents that govern your work as a SAS programmer in the pharmaceutical industry. Here are some of them:

- "E3 Structure and Content of Clinical Study Reports," found at http://www.fda.gov/cder/guidance/iche3.pdf.

- "E6 Good Clinical Practice: Consolidated Guidance," found at http://www.fda.gov/cder/guidance/959fnl.pdf.

- "E9 Statistical Principles for Clinical Trials," found at http://www.fda.gov/cder/guidance/ICH_E9-fnl.PDF.

- "Part 312.33 of Title 21 of the Code of Federal Regulations; Annual Reports," which dictates the content of the annual IND safety update, can be found at http://www.gpoaccess.gov/cfr/index.html.

- Current electronic submission regulations can be found at http://www.fda.gov/cder/regulatory/ersr/default.htm. This site includes the relevant documents "Providing Regulatory Submissions in Electronic Format – General Considerations," "Providing Regulatory Submissions in Electronic Format – NDAs," and the "Electronic Common Technical Document Specification."

Standards and Industry Organizations

Here is a list of industry and standards organizations that you should be familiar with.

- International Conference on Harmonization, at http://www.ifpma.org
- Clinical Data Interchange Standards Consortium (CDISC), at http://www.cdisc.org
- Society for Clinical Data Management, at http://www.scdm.org
- Association for Clinical Data Management, at http://www.acdm.org.uk/
- Drug Information Association, at http://www.diahome.org
- Applied Clinical Trials, at http://www.actmagazine.com
- Statisticians in the Pharmaceutical Industry (PSI), at http://www.psiweb.org. They have published the excellent "Guidelines for Standard Operating Procedures for Good Statistical Practice in Clinical Research," at http://psiweb.org/pdf/gsop.pdf.

SAS Help

In addition to help you can get from your coworkers, there is a vast amount of SAS programming support available to you. The following resources are available to help you enhance your SAS skills and to help solve your SAS problems.

SAS-L

SAS-L is an Internet mail list and USENET forum that has been active since the mid-1990s. It is the primary place on the Internet where SAS programmers share their problems, experiences, and expertise. If you have a SAS problem, it has probably already been discussed on SAS-L. You can subscribe, view, or post messages to the SAS-L listserv by going to http://www.listserv.uga.edu/cgi-bin/wa?A0=sas-l. I am fond of searching SAS-L through Google groups at http://groups.google.com/groups?hl=en&group=comp.soft-sys.sas.

SAS Technical Support

SAS provides excellent free unlimited technical support for all licensed SAS users. However, before involving SAS Technical Support, here is the process I would recommend when you have a SAS programming problem:

1. Take your problem to your company's SAS support personnel and resident SAS experts to see if they can help you.

2. Search the SAS Technical Support knowledge base at http://support.sas.com/techsup/search/index.html or search SAS-L at http://groups.google.com/groups?hl=en&group=comp.soft-sys.sas to see if there is a known solution to your problem.

3. Contact SAS Technical Support at http://support.sas.com/techsup/service_intro.html. You can discuss your problem with SAS by phone or by e-mail.

I recommend that you contact SAS Technical Support last primarily so that SAS can continue to offer this excellent service for free.

SAS Users Groups

SAS users groups are user-based communities supported by SAS Institute that are organized by geographical or industrial areas. These users groups typically have annual meetings where you can go to learn and to network with fellow SAS programmers.

Of particular interest to SAS users in the pharmaceutical industry is PharmaSUG, whose Web site can be found at http://www.pharmasug.org. In case you are unable to attend a SAS conference, both PharmaSUG and the SAS Users Group International (SUGI), at http://support.sas.com/usergroups/sugi/intro.html, make conference papers available on their Web sites after the conference. SAS maintains a list of all SAS users groups at http://support.sas.com/usergroups/.

SAS Manuals and Online Documentation

Hardcopy SAS manuals (user's guides and reference) and electronic documentation have evolved over the years. In SAS 6 and earlier versions, SAS manuals were available only in hardcopy, and you had to use the SAS windowing environment for electronic help. With SAS 8, you can still get hardcopy manuals and you can use the SAS windowing environment to access help just as before. However, SAS also sells a CD-ROM called *SAS OnlineDoc for SAS 8* that gives you the SAS manuals in HTML or PDF format. Finally, you can now get free SAS 8 documentation over the Internet in HTML format from SAS at http://support.sas.com/documentation/onlinedoc/index.html.

With the release of SAS 9, it is getting to the point where having a comprehensive set of hardcopy SAS manuals is difficult to afford in both monetary cost and physical storage space. The cost and size of SAS manuals have grown substantially. Also, each interim release of SAS now comes with a new set of documentation, so hardcopy versions of the manuals can rapidly get out of date. The SAS 9 manuals are provided by SAS Publishing on a "print on demand" basis, which means that SAS produces hardcopy SAS manuals as they are ordered, but once they have been ordered they are non-returnable and non-refundable. The good news is that the SAS manuals for SAS 9.0 and later versions can be found in HTML and PDF format online at http://support.sas.com/documentation/onlinedoc/index.html, so you can always view and print the current SAS manuals from there. Your challenge is to determine the best strategy for managing your SAS documentation.

SAS Press

SAS Press, previously known as SAS Books by Users Press, is a component of SAS Publishing that specializes in producing books written by the SAS user community. SAS Press books make an excellent supplement to the regular SAS manuals. Where the regular SAS manuals do a fine job of covering the capabilities of SAS software, SAS Press books go into further usage detail, offer many practical suggestions, and provide examples of industrial applications of SAS. There are dozens of books on a broad array of topics to choose from. For more information about SAS Press, go to http://support.sas.com/publishing/bbu/index.html.

Community Pages at SAS

The community Web pages at SAS go into further detail on the use and development of SAS software. These Web pages are constantly updated with helpful details on how to make SAS work for you and on future developments of SAS software. I have found the Base SAS community page to be especially useful. You can find the SAS community pages at http://support.sas.com/rnd/intro.html.

Third-Party SAS Web Pages

There are quite a few Web pages on the World Wide Web that address topics related to SAS. Many are organized in SAS Web rings, so that when you find one you can link to the others. Most of these Web pages are hosted by third-party SAS consultants and solution providers, but they can provide some good SAS tips and tricks. Simply do an Internet search on "SAS" to find these Web sites.

Useful Technical Skills

The world of information technology is full of tools that may aid you in becoming a better technology professional and SAS programmer. Here are just a few topics that you may want to explore further.

Scripting

You can use the *scripting language* of your computer operating system to automate many of the utility tasks that you need to perform. Some common useful scripts include SAS batch jobs submission and *make files*, SAS log error checking, project snapshot/backup, and reporting output integration. In the Microsoft environment you can write scripts in *BAT files*. In the UNIX or Linux environment you have several different *shells* that you can write scripts for. In the OpenVMS environment you can write *COM files* using *DCL*. Finally, there are scripting languages such as *Perl* that are cross-platform and extremely powerful for writing utility scripts. No matter what your SAS operating environment is, you will want to learn about scripting languages and use them to accomplish your various utility tasks.

Version Control Software

To keep control of your SAS programs you should learn about and invest in version control software. There are many commercial and free (*RCS* or *CVS*) version control software solutions available that do an excellent job of maintaining the integrity of your SAS programs. You can also integrate your version control software solution into your SAS program editor.

VBScript/JavaScript for Applications

Some software applications that you may use for reporting clinical trial information have their own scripting languages. For Microsoft Office applications you can write powerful local utility functions with *Visual Basic* or *VBScript*. Other applications, such as Adobe Acrobat, may use *JavaScript* for utility functions.

Systems Development Life Cycle

If you plan to get into serious applications development, or even if you plan to write comprehensive generic SAS macros, you will want to learn about the systems development life cycle (SDLC). Developed by the Institute of Electrical and Electronics Engineers (IEEE), the SDLC has been the traditional approach to software applications development for a long time. Besides the traditional SDLC, there are many derivative software development models that you may want to explore, such as Extreme Programming (XP) and Agile Development.

Modeling Tools

You may find modeling and visualization tools useful in building your SAS applications. The *Unified Modeling Language (UML)* is a language developed to help you map out your business processes and programming logic. UML can help you design your SAS application properly. Also, SAS Enterprise Guide 3.0 provides a SAS Process Flow window to visually show you the structure of your SAS application. Finally, there are third-party SAS program visualization tools available to help you see how your SAS programs and data interact.

Markup Languages

In the process of analysis and reporting for clinical trials, you may interact with various file markup languages. As you have already seen, XML is growing more prominent in clinical trial research as a means of data exchange. For reporting purposes, you may need to understand markup languages such as Rich Text Format (RTF) when working with Microsoft files, or HTML/XHTML when working with Web-based reports. The RTF documentation can be found on the Microsoft Developer Network Web site at http://msdn.microsoft.com/, and the HTML/XHTML/XML specifications can be found at the W3C Web site at http://www.w3.org.

File Transport and Data Encryption Technologies

You will want to understand the many ways to move SAS files electronically around the world. One means of file transport is File Transfer Protocol (FTP), which is built into many Web browsers, available at the command line, or available as a separate client program. A secure means of file transport is available with the Secure File Transport Protocol (SFTP). SFTP encrypts the electronic data stream being sent as well as the actual process of logging on to an SFTP server. Finally, before sending your data you will want to understand data encryption technologies. I recommend using PGP with key exchange for strong data security, but other file encryption tools are available as well. For instance, WinZip version 9 offers high-bit encryption schemes, although it does not use key exchange.

Other Applications Development Languages

If you are investigating applications development, I advise you to explore *Java* or *C#* as a possible programming language. Both Java and C# are strong object-oriented applications development languages with a rich set of classes and methods to help you develop a vast array of portable computer applications. Java has been around longer than C#, and you can find a considerable amount of documentation, training material, and the Java software itself at Sun Microsystem's Java site, at http://java.sun.com/. C# is the language Microsoft created to support its .NET Web services architecture. Because the majority of computers run a Microsoft operating system, writing applications that are compliant with the .NET architecture is a compelling idea.

Qualifying for and Obtaining a Job

How can you qualify for and obtain a SAS programming job in the pharmaceutical industry? SAS programmers, like all computer programmers, come from diverse backgrounds. The typical SAS programmer in the pharmaceutical industry has a science-based undergraduate degree and was exposed to SAS programming in college. Training to be a SAS programmer is usually done while on the job through work experience, company-sponsored classes, and SAS Institute courses. So if most of the training for these jobs is on the job, how can you position yourself to get a job as a SAS programmer in the pharmaceutical industry? Here are some things that will help you qualify for a job:

- Get a master's or bachelor's degree in statistics or computer science. Other science degrees that have a strong analytical or life science basis are useful as well.

- In lieu of experience, get certified. SAS offers a variety of certifications, including the relevant "SAS Certified Base Programmer" and "SAS Certified Advanced Programmer." The SAS certification Web page can be found at http://support.sas.com/certify/index.html. Philadelphia University offers a certification program called the "SAS Programming Certificate for the Pharmaceutical Industry." More information about this program can be found at http://www.philau.edu/continuinged/sascert.html. There are also other SAS certification programs that you can find by doing an Internet search.

- Network with professionals in the industry. Friends or contacts in the industry can serve as valuable job references. Attending local, regional, and national SAS users group meetings is an excellent way to network as well. SAS keeps a list of users groups at http://support.sas.com/usergroups/.

Glossary

adverse event

as defined by "E6 Good Clinical Practice: Consolidated Guidance," "any untoward medical occurrence in a patient or clinical investigation subject administered a pharmaceutical product and that does not necessarily have a causal relationship with this treatment. An AE can therefore be any unfavorable and unintended sign (including an abnormal laboratory finding), symptom, or disease temporally associated with the use of a medicinal (investigational) product, whether or not related to the medicinal (investigational) product." *See also* E6.

Analysis Dataset Model (ADaM)

as defined by CDISC, "a set of guidelines and examples for analysis datasets used to generate the statistical results for submission to a regulatory authority such as FDA. It specifically addresses needs of statistical reviewers." *See also* Clinical Data Interchange Standards Consortium.

annotated CRF

a case report form that has variable and data set names written on it to show where the data can be found in the clinical database. *See also* case report form.

application program interface (API)

a standard set of routines and software calls used to build computer programs. APIs make it easier to build software from modular blocks of code and easier for one program to speak to another program.

baseline

a measurement or assessment that occurs before an intervention of interest, such as dosing a patient with study therapy. In the event of multiple measurements before the intervention of interest, the baseline measurement is typically the last measurement that occurs before the intervention.

bias

> as defined by "E9 Statistical Principles for Clinical Trials," "the systematic tendency of any factors associated with the design, conduct, analysis and evaluation of the results of a clinical trial to make the estimate of a treatment effect deviate from its true value. Bias introduced through deviations in conduct is referred to as *operational bias*. The other sources of bias listed above are referred to as *statistical bias*." *See also* E9.

biologics

> defined by the FDA as products used to treat human disease that are derived or created from living organisms such as plants, animals, humans, or microorganisms.

blinding

> as defined by "E6 Good Clinical Practice: Consolidated Guidance," "a procedure in which one or more parties to the trial are kept unaware of the treatment assignment(s). Single blinding usually refers to the subject(s) being unaware, and double blinding usually refers to the subject(s), investigator(s), monitor, and, in some cases, data analyst(s) being unaware of the treatment assignment(s)." *See also* E6.

blocking factor

> a type of stratification within a randomization scheme that ensures that the randomization is balanced. Blocking factors ensure an even distribution of treatment assignments within any given strata. A block is typically a multiple of the treatment ratio and is usually large enough that an investigator cannot guess the next treatment. So, for example, if you have a treatment ratio of 1:1 of Active Drug to Placebo, you might have a block size of six or eight. A block size of two would be too small and would allow an investigator to guess that if a subject were just randomized to Placebo, the next patient would be randomized to Active Drug. *See also* randomization; randomization scheme; treatment ratio.

case report form (CRF)

> the assortment of paper forms used to collect a patient's clinical trial data for a given protocol. An individual page within a case report form is typically referred to as a "CRF page," while "case report form" most often is used to mean the collection of all CRF pages. When a paperless clinical data entry system is used, such as a Web-based electronic data capture system, you have what is called an "eCRF," or electronic case report form. *See also* clinical trial; electronic data capture; protocol.

case report tabulations (CRTs)

> listings of the clinical trial data by domain, such as demographics, adverse events, and medications. Case report tabulations are required for report submissions to the FDA. The CDISC Study Data Tabulation Model was designed to become an eventual replacement for case report tabulations in FDA report submissions. *See also* domain; Study Data Tabulation Model.

Clinical Data Interchange Standards Consortium (CDISC)

> as stated by CDISC, an organization whose mission is "to develop and support global, platform-independent data standards that enable information system interoperability to improve medical research and related areas of healthcare." CDISC is the driving force behind defining various data standards for the clinical trial industry. *See also* Analysis Dataset Model; Operational Data Model; Study Data Tabulation Model.

clinical data management system (CDMS)

> a relational data management system that focuses on the needs of managing clinical trial data. *See* relational database; relational data management system.

clinical endpoint committee (CEC)

> a group of clinicians who evaluate a patient's clinical data in order to determine whether an event of interest has occurred. Clinical endpoint committees are especially useful when an event determination has some subjective component. *See also* endpoint.

clinical investigator

> *See* investigator.

clinical narrative

> a detailed description of a patient's serious or significant adverse event or death. Usually the clinical narrative or narratives for a patient are preceded by a patient profile. Clinical narrative requirements are specified in the "E3 Structure and Content of Clinical Study Reports." *See also* E3; patient profile.

clinical site

> a doctor's office or clinic where a patient is seen during a clinical trial. The clinical site is responsible for conducting the clinical trial and provides the clinical source data for statistical analysis.

clinical study report (CSR)

as defined by "E6 Good Clinical Practice: Consolidated Guidance," "a written description of a trial/study of any therapeutic, prophylactic, or diagnostic agent conducted in human subjects, in which the clinical and statistical description, presentations, and analyses are fully integrated into a single report (see the ICH Guidance for Structure and Content of Clinical Study Reports)." *See also* E3; E6.

clinical trial

as defined by "E6 Good Clinical Practice: Consolidated Guidance," "any investigation in human subjects intended to discover or verify the clinical, pharmacological, and/or other pharmacodynamic effects of an investigational product(s), and/or to identify any adverse reactions to an investigational product(s), and/or to study absorption, distribution, metabolism, and excretion of an investigational product(s) with the object of ascertaining its safety and/or efficacy. The terms clinical trial and clinical study are synonymous." *See also* E6.

Common Technical Document (CTD)

an ICH-defined format for a regulatory submission that is considered acceptable in Japan, Europe, the United States, and Canada. *See also* International Conference on Harmonisation of Technical Requirements for Registration of Pharmaceuticals for Human Use.

concomitant medications

drugs taken by the patient during the course of the clinical trial that are not the active therapies under study. *See also* clinical trial.

contract research organization (CRO)

as defined by "E6 Good Clinical Practice: Consolidated Guidance," "a person or an organization (commercial, academic, or other) contracted by the sponsor to perform one or more of a sponsor's trial-related duties and functions." An academic CRO is often called an "academic research organization," or "ARO." *See also* E6.

covariate

a variable that has an influence on another dependent variable. For example, a history of smoking may be a covariate for whether or not a patient gets cancer.

crossover trial

> a clinical trial in which a patient receives both treatments to be compared. During the trial, the patient "crosses over" from one treatment to the other. *See also* clinical trial; parallel trial.

data clarification form (DCF)

> a document generated by clinical data management and sent to a clinical site to question/query a data value from the case report form so that it can be verified or corrected. *See also* case report form; clinical site; query.

data safety and monitoring board (DSMB)

> *See* Independent Data Monitoring Committee.

dependent variable

> a variable in which the response or value is determined in part by the response or value of other variables in an equation. *See also* independent variables.

domain

> as defined by the CDISC Study Data Tabulation Model, "a collection of observations with a topic-specific commonality about a subject." Demographics, adverse events, and concomitant medications are all examples of domains. *See also* Clinical Data Interchange Standards Consortium; Study Data Tabulation Model.

double blind trial

> a clinical trial is one in which the patient and the investigator are unaware of the actual study treatment the patient is receiving. *See also* clinical trial; single blind trial; triple blind trial.

drug assignment data set

> a data set that shows which patients were given which drug.

drug kit list

> a list that shows which drug container/kit label goes with which study medication. It sometimes needs to be referenced to verify drug assignments.

drug log

> the case report form that collects the details about study therapy dosing or intervention for an individual patient. *See also* case report form.

Dynamic Data Exchange (DDE)

a Microsoft communications protocol that allows two different software applications to send data and commands back and forth.

E-submission

electronic submission; refers to FDA regulatory drug submissions that are provided via computer files. The main components of current electronic submissions are linked PDF files and SAS transport format data sets. The future of electronic submissions will encompass the eCTD and XML data files. *See also* Electronic Common Technical Document; XML.

efficacy data

information about how a drug or device affects some disease state.

Electronic Common Technical Document (eCTD)

the electronic form of the Common Technical Document. *See also* Common Technical Document.

electronic data capture (EDC)

the technology that allows for clinical data to be obtained over a computer, alleviating the need for paper-based case report forms. Most electronic data capture systems today are Web-based computer applications.

endpoint

the key safety or efficacy event of concern in a clinical trial. An endpoint is typically the event the clinical trial was designed to study. *See also* efficacy data; safety data.

enroll

to consent to be a part of a clinical trial. Enrollment in a randomized clinical trial occurs before or at study therapy randomization. Patients may enroll in a clinical trial but never make it to study therapy randomization because they fail to meet certain protocol screening phase requirements. Therefore, the number of patients enrolled in a clinical trial should be as large or larger than the number of patients randomized in a trial. *See also* randomization.

equivalence trial

> as defined by "E9 Statistical Principles for Clinical Trials," "a trial with the primary objective of showing that the response to two or more treatments differs by an amount which is clinically unimportant. This is usually demonstrated by showing that the true treatment difference is likely to lie between a lower and an upper equivalence margin of clinically acceptable differences." *See also* E9.

events class

> a type of domain defined by the CDISC Study Data Tabulation Model that "captures occurrences or incidents independent of planned study evaluations occurring during the trial (e.g., 'adverse events' or 'disposition') or prior to the trial (e.g., 'medical history')." *See also* Clinical Data Interchange Standards Consortium; Study Data Tabulation Model.

E3

> shortened name for "E3 Structure and Content of Clinical Study Reports," which describes in detail what reporting goes into a clinical study report for an FDA submission. The E3 can be found at http://www.fda.gov/cder/guidance /iche3.pdf. *See also* clinical study report.

E6 (GCPs)

> shortened name for "E6 Good Clinical Practice: Consolidated Guidance," or GCPs; discusses the overall standards for implementing a clinical trial. It can be found at http://www.fda.gov/cder/guidance/959fnl.pdf.

E9

> shortened name for "E9 Statistical Principles for Clinical Trials"; discusses the statistical issues in the design and conduct of a clinical trial. It can be found at http://www.fda.gov/cder/guidance/ICH_E9-fnl.PDF.

Final Study Report (FSR)

> *See* clinical study report.

findings class

> a type of domain defined by the CDISC Study Data Tabulation Model that "captures the observations resulting from planned evaluations to address specific questions such as observations made during a physical examination, laboratory tests, ECG testing, and sets of individual questions listed on questionnaires." *See also* Clinical Data Interchange Standards Consortium; Study Data Tabulation Model.

Food and Drug Administration (FDA)

a division of the U.S. Department of Health and Human Services responsible for regulating the foods and drugs of U.S. consumers. The FDA states its mission as follows: "The FDA is responsible for protecting the public health by assuring the safety, efficacy, and security of human and veterinary drugs, biological products, medical devices, our nation's food supply, cosmetics, and products that emit radiation. The FDA is also responsible for advancing the public health by helping to speed innovations that make medicines and foods more effective, safer, and more affordable; and helping the public get the accurate, science-based information they need to use medicines and foods to improve their health."

free-text variable

a character SAS variable that can contain any ASCII text whatsoever. The variable is called "free" because there are no restrictions on the content of the variable when it is entered. Free-text variables by themselves are of limited use in statistical analysis. A free-text field usually needs to be categorized in order to be of use statistically.

FTP

File Transfer Protocol. A protocol used to move files over the Internet. Typically, you use a command line command, browser capability, or program on your computer called an FTP client to connect to an FTP server found on the Internet. Once connected to the FTP server, you can move files back and forth between your computer and the FTP server. SFTP, Secure FTP, is a more secure protocol for moving data over the Internet in which both your login information and data are encrypted during communications.

hardcode

a line of code in a computer program that causes an action to be taken based on an explicit data value. The action may include changing the content of a variable, deleting an observation, or some other action. The following is an example of a hardcode:

```
**** HARDCODE DISCONTINUATION = DEATH FOR SUBJECT 101-1002;
data endstudy;
   set endstudy;

   if subjid = "101-1002" then
      discterm = "Death";
run;
```

HTML (Hypertext Markup Language)

defined by the World Wide Web Consortium (W3C) as "the lingua franca for publishing hypertext on the World Wide Web. It is a non-proprietary format based upon SGML, and can be created and processed by a wide range of tools, from simple plain text editors—you type it in from scratch—to sophisticated WYSIWYG authoring tools. HTML uses tags such as <h1> and </h1> to structure text into headings, paragraphs, lists, hypertext links, etc." *See also* SGML; WYSIWYG.

Independent Data Monitoring Committee (IDMC)

as defined by "E6 Good Clinical Practice: Consolidated Guidance," a group "that may be established by the sponsor to assess at intervals the progress of a clinical trial, the safety data, and the critical efficacy endpoints, and to recommend to the sponsor whether to continue, modify, or stop a trial." An IDMC is sometimes referred to as a data safety and monitoring board when patient safety is of the primary concern. *See also* E6.

independent variables

variables in which the response or value helps to determine the value of other variables in an equation. *See also* dependent variable.

inferential analyses

statistical analyses on a sample taken from a larger study population that result in being able to make generalizations and conclusions about the larger study population.

instance (Oracle)

the shared memory and process area that govern a set of Oracle databases. Many users, processes, and databases may share a given Oracle instance.

Integrated Clinical/Statistical Report

See clinical study report.

Interactive Voice Response System (IVRS)

a telephone voice-activated computer system often used for randomization assignment. In practice, a site calls into the IVRS and requests a drug assignment, and the IVRS assigns the next appropriate treatment from the randomization scheme. *See also* randomization; randomization scheme; clinical site.

International Conference on Harmonisation of Technical Requirements for Registration of Pharmaceuticals for Human Use (ICH)

> defined by the ICH as "a unique project that brings together the regulatory authorities of Europe, Japan and the United States and experts from the pharmaceutical industry in the three regions to discuss scientific and technical aspects of product registration. The purpose is to make recommendations on ways to achieve greater harmonisation in the interpretation and application of technical guidelines and requirements for product registration in order to reduce or obviate the need to duplicate the testing carried out during the research and development of new medicines."

intervention class

> a type of domain defined by the CDISC Study Data Tabulation Model that "captures investigational treatments, therapeutic treatments, and surgical procedures that are intentionally administered to the subject (usually for therapeutic purposes) either as specified by the study protocol (e.g., 'exposure') or coincident to the study assessment period (e.g., 'concomitant medications')." *See also* Clinical Data Interchange Standards Consortium; Study Data Tabulation Model.

Investigational New Drug (IND) application

> as defined by the FDA, "an application that a drug sponsor must submit to FDA before beginning tests of a new drug on humans. The IND contains the plan for the study and is supposed to give a complete picture of the drug, including its structural formula, animal test results, and manufacturing information."

investigator

> as defined by "E6 Good Clinical Practice: Consolidated Guidance," "a person responsible for the conduct of the clinical trial at a trial site. If a trial is conducted by a team of individuals at a trial site, the investigator is the responsible leader of the team and may be called the principal investigator." *See also* E6; clinical site.

laboratory data

> a set of measurements of a patient usually involving urinalysis, hematology, and blood chemistry tests but possibly also ECG, microbiologic, and other therapeutic-indication-specific clinical lab tests.

laboratory normal range

a set of values that define what an expected lab value should be for an average patient in a clinical trial. Typically, there is a low normal value and a high normal value for a laboratory parameter. Some clinical trials include "panic" or "level of concern" laboratory ranges, which are more extreme than laboratory normal ranges. *See also* laboratory data.

markup tags

text flags inserted into a text data stream to indicate structure and content of the data stream. XML, HTML, and SGML all use markup tags to define their respective languages. *See also* HTML; SGML; XML.

MedDRA

Medical Dictionary for Regulatory Activities; a creation of the International Conference on Harmonisation used to categorize and code diseases, disorders, and adverse events.

medical history

information about a patient's previous medical conditions. Data about previous conditions with special relevance to the indication under study are especially important to collect.

multi-center trial

as defined by "E6 Good Clinical Practice: Consolidated Guidance," "a clinical trial conducted according to a single protocol but at more than one site, and, therefore, carried out by more than one investigator." *See also* clinical trial; E6.

New Drug Application (NDA)

as defined by the FDA, "an application requesting FDA approval to market a new drug for human use in interstate commerce. The application must contain, among other things, data from specific technical viewpoints for FDA review— including chemistry, pharmacology, medical, biopharmaceutics, statistics, and, for anti-infectives, microbiology."

Object Linking and Embedding (OLE)

a protocol written by Microsoft that allows the passing of objects between computer applications.

Open Database Connectivity (ODBC)

an open and independent software applications program interface for connecting to databases such as the data in the clinical data management system. *See also* applications program interface; clinical data management system.

Operational Data Model (ODM)

as defined by CDISC, "a vendor neutral, platform independent format for interchange and archive of data collected from various sources in clinical trials. The model represents study metadata, clinical data and administrative data associated with a clinical trial. Information that needs to be shared among different software systems during a trial, or archived after a trial, is included in the model. The model complies with FDA 21CFR11 regulations." *See also* Clinical Data Interchange Standards Consortium.

output shells

See table shell.

parallel trial

a clinical trial in which a patient receives only one study therapy for the duration of the trial. Patients who receive other therapies are said to be running in parallel to the patients on the other treatment arm. *See also* clinical trial; crossover trial.

parent-child

a set of variables in which a lower-order variable should be answered only when the higher-order variable is answered in affirmation of the subordinate variable. Parent-child variables are the source of numerous data management queries in paper-based case report forms. *See also* case report form; query.

patient profile

a comprehensive description of a patient at a given point in time. The patient profile often contains the patient's demographics, medical history, physical exam results, and current medications. Graphics may be provided for data such as laboratory data values. A patient profile often precedes a clinical narrative for a patient in a clinical study report. *See also* clinical narrative.

prior medications

drugs taken by a patient before enrollment in a clinical trial. *See also* clinical trial; enroll.

protocol

as defined by "E6 Good Clinical Practice: Consolidated Guidance," "a document that describes the objective(s), design, methodology, statistical considerations, and organization of a trial. The protocol usually also gives the background and rationale for the trial, but these could be provided in other protocol referenced documents." *See also* E6.

quality-of-life (QOL) data

data that encompass traditional information about a patient's physical wellbeing along with mental, social, and economic health measures. There are numerous quality-of-life measurement scales that attempt to measure the overall wellbeing of a patient. Clinical trials sometimes gather quality-of-life measures as secondary endpoint measures. *See also* clinical trial; endpoint.

query

a question asked by clinical data management about the clinical data found on the case report form. A query usually results in a data clarification form being sent to the site. *See also* case report form; clinical site; data clarification form.

randomization

as defined by "E6 Good Clinical Practice: Consolidated Guidance," "the process of assigning trial subjects to treatment or control groups using an element of chance to determine the assignments in order to reduce bias." *See also* enroll; E6.

randomization scheme

an ordered list of study drug assignments. Strata, such as site, typically break up the randomization scheme, but within each stratum the order of drug assignments is critical to maintaining the randomization scheme. *See also* randomization.

relational database

a set of normalized data tables that can have rows associated and linked with one another through the use of key variables in the tables.

relational database management system (RDBMS)

a software program that maintains a relational database. *See also* relational database.

safety data

information about whether a drug or device is harmful to a patient. The FDA states this about safety: "No drug is completely safe or without the potential for side effects. Before a drug may be approved for marketing, the law requires the submission of results of tests adequate to show the drug is safe under the conditions of use in the proposed labeling. Thus, 'safety' is determined case by case and reflects the drug's risk-vs.-benefit relationship."

serious adverse event (SAE)

as defined by "E6 Good Clinical Practice: Consolidated Guidance," "any untoward medical occurrence that at any dose: results in death, is life-threatening, requires inpatient hospitalization or prolongation of existing hospitalization, results in persistent or significant disability/incapacity, or is a congenital anomaly/birth defect." A serious adverse event is sometimes confused with an adverse event that is marked as "severe" on the case report form. Although a serious adverse event may be severe, a severe adverse event does not have to be serious. *See also* E6.

SGML (Standardized Generalized Markup Language)

as defined by the World Wide Web Consortium (W3C), "a system for defining markup languages. Authors mark up their documents by representing structural, presentational, and semantic information alongside content. HTML is one example of a markup language." *See also* HTML.

single blind trial

a clinical trial in which only the patient is unaware of the actual study treatment being given to patients. *See also* clinical trial; double blind trial; triple blind trial.

site

See clinical site.

statistical analysis plan (SAP)

a comprehensive document that describes how the clinical trial data will be analyzed to comply with the directives of the clinical protocol. Whereas the protocol has a paragraph or two describing the statistical analysis, the statistical analysis plan describes in detail the inferential analyses to be done, and includes study population definitions, data windowing or other special data handling rules, and often draft output shells that show precisely what tables, listings, and graphs will be provided in the reporting.

structured query language (SQL)

the computer language most often used to query and manipulate data from a relational database. *See also* relational database.

Study Data Tabulation Model (SDTM)

as defined by CDISC, a "set of standards developed by CDISC that are intended to guide the organization, structure, and format of standard clinical trial tabulation datasets submitted to a regulatory authority such as the US Food and Drug Administration (FDA). This model is specified in FDA Guidance for implementation of the eCTD." See also case report tabulations; Clinical Data Interchange Standards Consortium; Electronic Common Technical Document.

study day

the number of days that some event occurs after study therapy intervention. Events that occur prior to randomization typically have negative study day values. Some define the day before study therapy intervention as study day 0 (CDISC ADaM models), while others define the day before study therapy intervention as study day –1 (CDISC SDTM model). See also Analysis Dataset Model; Clinical Data Interchange Standards Consortium; Study Data Tabulation Model.

superiority trial

as defined by "E9 Statistical Principles for Clinical Trials," "a trial with the primary objective of showing that the response to the investigational product is superior to [the response to] a comparative agent (active or placebo control)." *See also* E9.

table shell

an outline or sketch, often found in the statistical analysis plan (SAP), of what a statistical table will look like in the completed reporting. The table shell shows how the table will look and contains row and column definitions for a statistical table. The table shell is often annotated with database variable names so that it can serve as a specification for the statistical programmer. There may be supplemental instructions for the programmer for complicated analysis tables.

TFL

an abbreviation for "tables, figures, and listings" (also TLG, for "tables, listings, and graphs"). Tables, figures and listings describe the types of reporting that a statistical programmer will produce for a clinical trial.

treatment-emergent signs and symptoms (TESS)

as defined by the FDA, "events not seen at baseline and events that worsened even if present at baseline."

treatment ratio

the comparison or numerical relationship between study therapies in a randomization scheme. Most clinical trials have equal balanced treatment assignments, so in the case of Active Drug versus Placebo, the treatment ratio would be 1:1, or "one to one." For each person who gets assigned Active Drug, another person gets assigned Placebo.

triple blind trial

a clinical trial in which the patient, the investigator, and the staff performing the data analysis are unaware of the actual study treatment being given to the patient. *See also* clinical trial; double blind trial; single blind trial.

21 CFR – Part 11

sometimes referred to as "Part 11"; a federal law that regulates the submission of electronic records and electronic signatures to the FDA. Of particular interest to statistical programmers is the requirement for computer systems validation. The full law can be found at http://www.fda.gov/ora/compliance_ref/part11/FRs/background/pt11finr.pdf.

21 CFR - Part 312.33

a federal law that discusses what is required for an investigational new drug application (IND). Of particular importance to a statistical programmer are the requirements for the annual reporting for the IND. This document can be found at http://www.gpoaccess.gov/cfr/index.html.

WHODrug

World Health Organization Drug Dictionary. A creation of the Uppsala Monitoring Centre in Sweden designed to take international, proprietary, and nonproprietary or generic drug names and classify them into common preferred terms.

WYSIWYG

pronounced "whizzy-wig"; stands for What You See Is What You Get. WYSIWYG means that what you see on your computer screen is what you get when you print or export that information. Depending on your printer drivers, Microsoft Word files may not be WYSIWYG. Also, Microsoft Excel files may not be WYSIWYG when they are exported to SAS data sets.

XML (Extensible Markup Language)

as defined by the World Wide Web Consortium (W3C), "a simple, very flexible text format derived from SGML (ISO 8879). Originally designed to meet the challenges of large-scale electronic publishing, XML is also playing an increasingly important role in the exchange of a wide variety of data on the Web and elsewhere." *See also* SGML.

Index

Visual Basic programming 299

W

Web-based community pages 298
WHODrug dictionary 111–112
Wilcoxon rank sum test 257
Wilcoxon signed rank test 145, 256
windowing data 91–94
work process 2
working environment of statistical
          programmers 10–18, 290
World Wide Web 290

X

XML files
    creating 266–275
    importing 68–79
XML LIBNAME engine
    creating XML files 274–275
    importing XML files 69–72
XML Mapper 72–73
XPORT transport format 264–265

Y

Y2K problem 112–114
YEARCUTOFF system option 112–114

Z

zero results vs. missing values 102–106

Numerics

21 CFR – Part 11 law 6, 25, 291, 294
21 CFR – Part 312.33 law 7, 295
2x2 tests for association 251–252

Books Available from SAS® Press

Advanced Log-Linear Models Using SAS®
by **Daniel Zelterman**

Analysis of Clinical Trials Using SAS®: A Practical Guide
by **Alex Dmitrienko, Geert Molenberghs, Walter Offen,** and **Christy Chuang-Stein**

Annotate: Simply the Basics
by **Art Carpenter**

Applied Multivariate Statistics with SAS® Software, Second Edition
by **Ravindra Khattree** and **Dayanand N. Naik**

Applied Statistics and the SAS® Programming Language, Fourth Edition
by **Ronald P. Cody** and **Jeffrey K. Smith**

An Array of Challenges — Test Your SAS® Skills
by **Robert Virgile**

Carpenter's Complete Guide to the SAS® Macro Language, Second Edition
by **Art Carpenter**

The Cartoon Guide to Statistics
by **Larry Gonick** and **Woollcott Smith**

Categorical Data Analysis Using the SAS® System, Second Edition
by **Maura E. Stokes, Charles S. Davis,** and **Gary G. Koch**

Cody's Data Cleaning Techniques Using SAS® Software
by **Ron Cody**

Common Statistical Methods for Clinical Research with SAS® Examples, Second Edition
by **Glenn A. Walker**

Debugging SAS® Programs: A Handbook of Tools and Techniques
by **Michele M. Burlew**

Efficiency: Improving the Performance of Your SAS® Applications
by **Robert Virgile**

The Essential PROC SQL Handbook for SAS® Users
by **Katherine Prairie**

Genetic Analysis of Complex Traits Using SAS®
by **Arnold M. Saxton**

A Handbook of Statistical Analyses Using SAS®, Second Edition
by **B.S. Everitt** and **G. Der**

Health Care Data and SAS®
by **Marge Scerbo, Craig Dickstein,** and **Alan Wilson**

The How-To Book for SAS/GRAPH® Software
by **Thomas Miron**

support.sas.com/pubs

Using the SAS® Windowing Environment:
A Quick Tutorial
by **Larry Hatcher**

Visualizing Categorical Data
by **Michael Friendly**

Web Development with SAS® by Example
by **Frederick Pratter**

Your Guide to Survey Research Using the
SAS® System
by **Archer Gravely**

JMP® Books

JMP® for Basic Univariate and Multivariate Statistics:
A Step-by-Step Guide
by **Ann Lehman, Norm O'Rourke, Larry Hatcher,**
and **Edward J. Stepanski**

JMP® Start Statistics, Third Edition
by **John Sall, Ann Lehman,**
and **Lee Creighton**

Regression Using JMP®
by **Rudolf J. Freund, Ramon C. Littell,**
and **Lee Creighton**

support.sas.com/pubs